SPRINGER PROTOCOLS HANDBOOKS

For further volumes:
http://www.springer.com/series/8623

Springer Protocols Handbooks collects a diverse range of step-by-step laboratory methods and protocols from across the life and biomedical sciences. Each protocol is provided in the Springer Protocol format: readily-reproducible in a step-by-step fashion. Each protocol opens with an introductory overview, a list of the materials and reagents needed to complete the experiment, and is followed by a detailed procedure supported by a helpful notes section offering tips and tricks of the trade as well as troubleshooting advice. With a focus on large comprehensive protocol collections and an international authorship, Springer Protocols Handbooks are a valuable addition to the laboratory.

DNA Modification Detection Methods

Edited by

Bi-Feng Yuan

Department of Chemistry, Wuhan University, Wuhan, China

Humana Press

Editor
Bi-Feng Yuan
Department of Chemistry
Wuhan University
Wuhan, China

ISSN 1949-2448 ISSN 1949-2456 (electronic)
Springer Protocols Handbooks
ISBN 978-1-0716-1228-6 ISBN 978-1-0716-1229-3 (eBook)
https://doi.org/10.1007/978-1-0716-1229-3

This Humana imprint is published by the registered company Springer Science+Business Media, LLC part of Springer Nature.
The registered company address is: 1 New York Plaza, New York, NY 10004, U.S.A.

Contents

PART III DETECTION OF 5HMU AND 5FU

PART IV DETECTION OF BASE J AND 8-OXO-7,8-DIHYDROGUANINE

Part I

Detection of 5mC, 5hmC, 5fC, and 5caC

Chapter 1

Quantitative Assessment of the Oxidation Products of 5-Methylcytosine in DNA by Liquid Chromatography-Tandem Mass Spectrometry

Jiekai Yin, Shuo Liu, and Yinsheng Wang

Abstract

Methylation at the C5 position of cytosine in DNA constitutes a prominent epigenetic modification. Reactive oxygen species (ROS) are ubiquitous within cells and they are very reactive toward 5-methyl-2′-deoxycytidine (5-mdC). Additionally, previous studies showed that 5-mdC could also be oxidized by ten-eleven translocation (Tet) family dioxygenases. Quantitative measurement of oxidation products of 5-mdC can facilitate the investigation about the roles of these modifications in gene regulation. Here we describe an LC-MS/MS/MS-based method for the sensitive and accurate quantifications of 5-hydroxymethyl-2′-deoxycytidine (5-HmdC), 5-formyl-2′-deoxycytidine (5-FodC), 5-carboxyl-2′-deoxycytidine (5-CadC), and 5-hydroxymethyl-2′-deoxyuridine (5-HmdU) in genomic DNA.

Keywords DNA modifications, DNA methylation, Epigenetics, Liquid chromatography-mass spectrometry

1 Introduction

Genetic information is embedded in the DNA sequence and the expression of genes can be regulated by DNA and histone modifications [1, 2]. In mammals, methylation of DNA at the C5 position of cytosine residues occurs primarily at CpG dinucleotide sites and plays an important role in the epigenetic mechanisms of gene regulation [3].

Reactive oxygen species (ROS) are ubiquitous in cells and can be continually generated from metabolic activity and/or upon exposure to environmental agents [4, 5]. ROS comprise of superoxide anion radical ($O_2^{-}\bullet$), hydroxyl radical (•OH), and hydrogen peroxide (H_2O_2), and •OH can abstract a hydrogen atom from 5-methyl-2′-deoxycytidine (5-mdC) [6, 7]. The resulting methyl radical can react with molecular oxygen to give 5-hydroxymethyl-2′-deoxycytidine (5-HmdC) and 5-formyl-2′-deoxycytidine (5-FodC).

Bi-Feng Yuan (ed.), *DNA Modification Detection Methods*, Springer Protocols Handbooks,
https://doi.org/10.1007/978-1-0716-1229-3_1,

5-mdC can also be oxidized by ten-eleven translocation (Tet) family dioxygenases in mammals [8, 9]. Tet1 was the first enzyme found to oxidize 5-mdC to 5-HmdC [9]. Later, all three members of the Tet family were shown to exhibit the same enzymatic activity. Tet can further oxidize 5-HmdC to yield 5-formyl-2-′-deoxycytidine (5-FodC) and 5-carboxyl-2′-deoxycytidine (5-CadC) (Fig. 1) [10, 11].

Apart from being an oxidation product of thymidine, 5-hydroxymethyl-2′-deoxyuridine (5-HmdU) may also be generated from the deamination of 5-HmdC [12, 13]. It was suggested that the deamination is facilitated by AID (activation-induced deaminase)/APOBEC (apolipoprotein B mRNA-editing enzyme complex) families of cytidine deaminases, though later biochemical studies did not support this notion (Fig. 1) [14].

5-FodC, 5-CadC, and 5-HmdU could be directly excised by thymine DNA glycosylase (TDG) [12, 13]. Subsequently, an unmethylated 2′-deoxycytidine would be restored by base-excision repair (BER) pathway. This provides new insight about active DNA cytosine demethylation in mammals (Fig. 1).

Quantitative measurements of oxidation products of 5-mdC can help explore the roles these modifications play in gene regulation. In this chapter, we describe a protocol for accurate and sensitive measurements of 5-HmdC, 5-FodC, 5-CadC, and 5-HmdU, which is based on a previously published study [15].

2 Materials

All chemicals and enzymes, unless otherwise specified, can be obtained from Sigma-Aldrich (St. Louis, MO). And all the solutions should be prepared with Milli-Q water (Milli-Q apparatus, Millipore).

2.1 DNA Extraction

2.1.1 Enzymes

1. Proteinase K: New England Biolabs (Ipswich, WA).
2. RNase A: 10 mg/mL.
3. RNase T1: 25 units/μL.

2.1.2 Chemicals and Solutions

1. Ethanol.
2. Chloroform/isoamyl alcohol (24:1, v/v).
3. Lysis buffer: 20 mM Tris (pH 8.1), 20 mM EDTA, 400 mM NaCl, 1% SDS (w/v).
4. Saturated sodium chloride solution.

2.1.3 Equipment

1. Nano-drop.

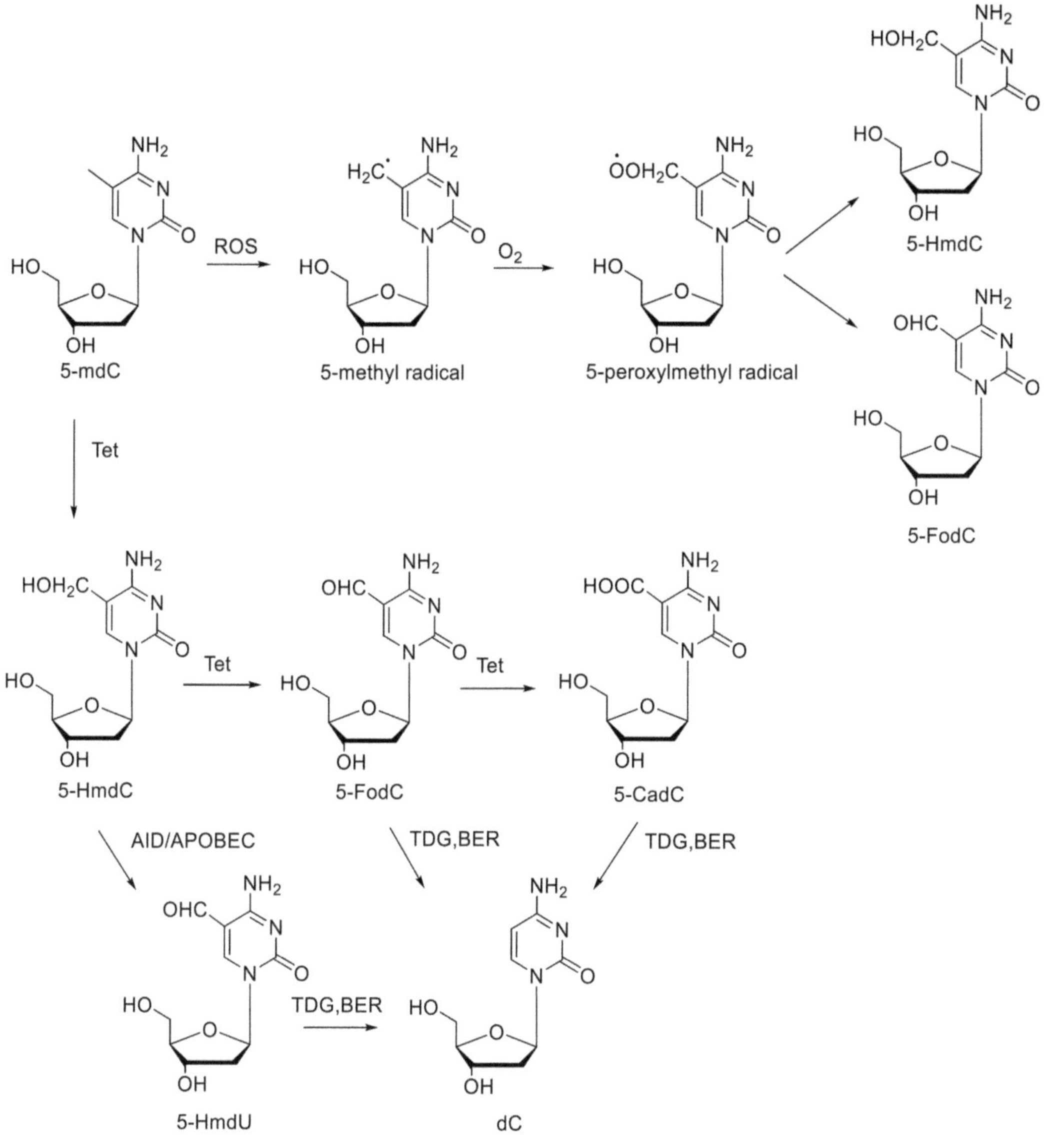

Fig. 1 Oxidation of 5-mdC in DNA and active DNA demethylation

2.2 Enzymatic Digestion

2.2.1 Enzymes

1. Nuclease P1.
2. Alkaline phosphatase.
3. Phosphodiesterase 1.
4. Phosphodiesterase 2.

2.2.2 Chemicals and Solutions

1. *erythro*-9-(2-hydroxy-3-nonyl)adenine (EHNA).
2. Formic acid.
3. 10 × Buffer I: 300 mM sodium acetate and 10 mM zinc acetate (pH 5.6).

4. 10 × Buffer II: 5 M Tris-HCl buffer (pH 8.9).

2.3 HPLC Enrichment

2.3.1 Chemicals and Solutions

1. 10 mM ammonium formate.
2. Methanol.

2.3.2 Equipment

1. A Beckman HPLC system with pump module 125 and UV detector module 126.
2. An Aeris Widepore C18 column (4.6 × 250 mm, 3.6μm in particle size, Phenomenex, Torrance, CA).
3. Speed-vac.

2.4 Quantification with LC-MS/MS/MS

2.4.1 Chemicals and Solutions

1. 0.10% (v/v) formic acid in water.
2. 0.10% (v/v) formic acid in methanol.
3. 2.0 mM ammonium formate in water.
4. 2.0 mM ammonium formate in methanol.

2.4.2 Equipment

1. An LTQ linear ion trap mass spectrometer (Thermo Fisher Scientific) coupled with an Agilent 1200 capillary HPLC pump (Agilent Technologies, Santa Clara, CA) was used to perform the LC-MS/MS/MS quantification of 5-HmdC, 5-FodC, and 5-CadC in the positive-ion mode, and 5-HmdU in the negative-ion mode.
2. A 0.5 × 250 mm Zorbax SB-C18 column (5μm in particle size, Agilent Technologies, Santa Clara, CA).

3 Method

3.1 DNA Extraction

Genomic DNA was extracted from cells by using a high-salt method [16].

1. Turn on a water bath and set the temperature to 55 °C.
2. Add 50μL of lysis buffer containing 20 mM Tris (pH 8.1), 20 mM EDTA, 400 mM NaCl, 1% SDS (w/v) as well as 10μL of proteinase K (20 mg/mL) to the cell pellet. Mix the sample by vortexing (*see* **Note 1**).
3. Incubate the sample in a 55 °C water bath overnight. If possible, vortex the samples occasionally.
4. Add saturated sodium chloride solution (0.5 volume) to the digestion mixture. Vortex the mixture for at least 1 min and incubate it at 55 °C for another 15 min (*see* **Note 2**).
5. Centrifuge the sample at 16,000 × g for 30 min at 4 °C, remove supernatant to a new labeled 1.5 mL microcentrifuge tube.

6. Add 2 volumes of cold 100% ethanol and gently mix by inverting the tube a couple of times; DNA precipitates out of solution under this condition (*see* **Note 3**).
7. Centrifuge the resulting mixture at 16,000 × g for 10 min at 4 °C. Discard the supernatant. Add 100μL of cold 70% ethanol to wash the DNA pellet. Centrifuge the sample at 16,000 × g for 5 min at 4 °C. Discard the supernatant and dissolve the DNA pellet in 95μL Milli-Q H_2O.
8. Add 3μL of RNase A (10 mg/mL) and 2μL of RNase T1 (25 units/μL) to the solution. Incubate the sample at 37 °C overnight (*see* **Note 4**).
9. Increase the volume of the sample to 300–600μL. Add an equal volume of chloroform/isoamyl alcohol (24:1, v/v) and mix by vortexing for a brief period of time. Centrifuge the resulting mixture at 16,000 × g for 10 min at room temperature. Transfer the aqueous layer to a new labeled 1.5 mL microcentrifuge tube. Decrease the volume to 30μL by Speed-vac.
10. Add 2 volumes of cold 100% ethanol and gently mix by inverting the tube a couple of times, which leads to the precipitation of DNA out of solution.
11. Centrifuge the sample at 16,000 × g for 10 min at 4 °C. Discard the supernatant. Add 100μL of cold 70% ethanol to wash the DNA pellet. Centrifuge the sample at 16,000 × g for 5 min at 4 °C. Discard the supernatant. Allow the sample to air dry at room temperature, then dissolve it in Milli-Q water, and quantify the sample by Nano-drop.

3.2 Enzymatic Digestion

1. Add 0.1 unit of nuclease P1, 0.000125 unit of phosphodiesterase 2, and 2.5 nmol of EHNA to 1μg cellular DNA in a solution containing 30 mM sodium acetate (pH 5.6) and 1.0 mM zinc acetate. Incubate the mixture at 37 °C for 48 h.
2. Add 0.1 unit of alkaline phosphatase, 0.00025 unit of phosphodiesterase 1, and 500 mM Tris-HCl buffer (pH 8.9) to the mixture. Incubate the mixture at 37 °C for another 2 h.
3. Neutralize the digestion mixture with 1.0 M formic acid.
4. For LC-MS/MS/MS quantification of 5-HmdC, add 50 fmol of [1,3-$^{15}N_2$-2′-D]-5-HmdC to the enzymatic digestion mixture of 10–50 ng genomic DNA. Remove the enzyme by chloroform extraction. The aqueous layer should be direct for the LC-MS/MS/MS analysis (*see* **Note 5**).
5. For 5-FodC, 5-CadC, and 5-HmdU, add 30 fmol of [1,3-$^{15}N_2$-2′-D]-5-FodC, 25 fmol of [4-amino-1,3-$^{15}N_3$]-5-CadC, and 2 pmol of [1,3-$^{15}N_2$-2′-D]-5-HmdU to the mixture. Dry the aqueous layer by Speed-vac, reconstitute it with

Milli-Q water, and inject it to offline HPLC for the enrichment of target nucleosides.

3.3 HPLC Enrichment

1. Carry out the HPLC enrichment on a Beckman HPLC system with pump module 125 and UV detector module 126.
2. To enrich the oxidative products of 5-mdC in the nucleoside mixture, an Aeris Widepore C18 column (4.6 × 250 mm, 3.6μm in particle size, Phenomenex, Torrance, CA) was used.
3. Use the 10 mM ammonium formate (pH 8.5) as the mobile phase and set the flow rate at 0.8 mL/min. A representative HPLC trace for the enrichment of 5-CadC, 5-HmdU, and 5-FodC is shown in Fig. 2.
4. Collect the HPLC fractions containing 5-CadC (3.0–4.0 min), 5-HmdU (9.5–10.0 min), and 5-FodC (26.0–27.5 min) individually and dry them in a Speed-vac.
5. Reconstitute the dried fractions in Milli-Q water, and inject the samples for LC-MS/MS/MS analysis (*see* **Note 6**).

3.4 Quantification with LC-MS/MS/MS

1. An LTQ linear ion trap mass spectrometer (Thermo Fisher Scientific) coupled with an Agilent 1200 capillary HPLC pump can be used to perform the LC-MS/MS/MS quantification of 5-HmdC, 5-FodC, and 5-CadC in the positive-ion mode, and 5-HmdU in the negative-ion mode.
2. Conduct the HPLC separation with a 0.5 × 250 mm Zorbax SB-C18 column (5μm in particle size, Agilent Technologies) at a flow rate of 8.0μL/min.
3. Use a solution of 0.1% (v/v) formic acid in water as mobile phase A and a solution of 0.1% (v/v) formic acid in methanol as mobile phase B for 5-HmdC, 5-FodC, and 5-CadC analyses. Use a solution of 2 mM ammonium formate in water as solution A and a solution of 2 mM ammonium formate in methanol as solution B for 5-HmdU analysis.
4. Use a gradient (5 min 0–20% B and 25 min 20–70% B) for all the modified nucleosides.
5. Use the optimized working parameters of the LTQ mass spectrometer, i.e. normalized collision energy and activation Q, for the sensitive detection of these modified nucleosides (Table 1) (*see* **Note 7**).

3.5 Data Analysis

1. Based on the peak area ratio in the selected-ion chromatograms for the analytes over their corresponding isotope-labeled standards, the amounts of the labeled standards added and the equations derived from the calibration curves, the moles of 5-HmdC, 5-FodC, 5-CadC, and 5-HmdU in the nucleoside mixtures can be calculated.

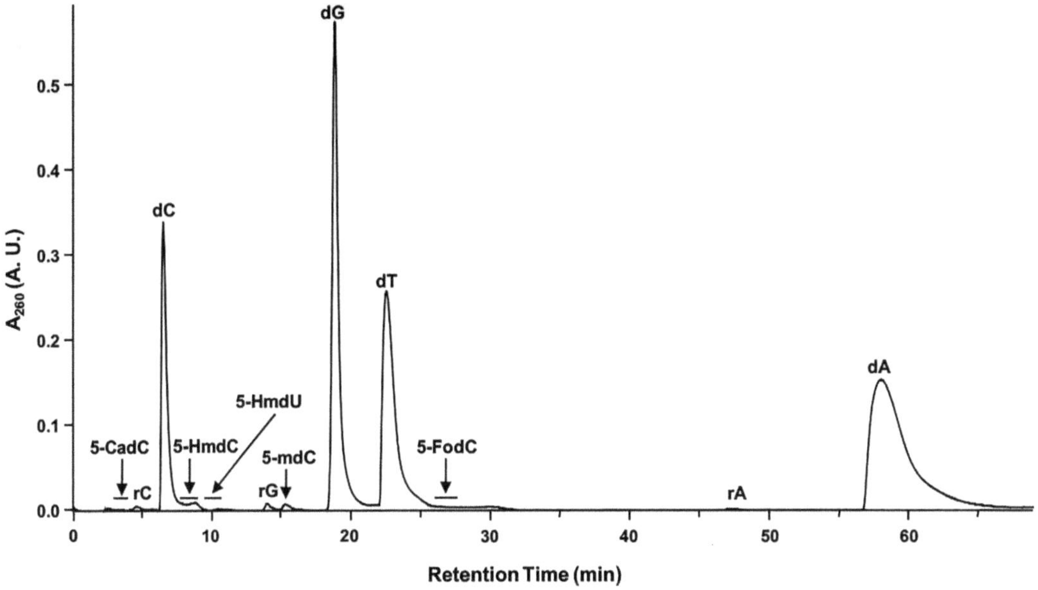

Fig. 2 HPLC trace for the enrichment of 5-CadC, 5-HmdU, and 5-FodC. (Reproduced from ref. [15] with permission from Oxford University Press)

Table 1
Optimized parameters for the detection of 5-HmdC, 5-FodC, 5-CadC, and 5-HmdU. (Reproduced from ref. [15] with permission from Oxford University Press)

		m/z transitions for selected-ion monitoring (MS^3)			
Analytes	**Ion mode**	**Unlabeled analyte**	**Isotope-labeled standard**	**Normalized collision energy (%)**	**Activation Q**
5-HmdU	–	257→214	260→216	30	0.25
		214→124	216→124	33	0.25
5-HmdC	+	258→142	261→145	48	0.25
		142→124	145→126	27	0.25
5-FodC	+	256→140	259→142	33	0.27
		140→97	142→98	36	0.43
5-CadC	+	272→156	275→159	15	0.3
		156→138	159→141	21	0.37

2. The number of moles of the modified nucleosides are divided by the number of moles of total nucleosides (or 5-mdC) in the digested DNA mixture to give the final DNA modification levels, which are expressed in terms of numbers of modifications per 10^6 2′-deoxynucleosides (or per 10^3 5-mdC). In this respect, the level of RNA contamination in the DNA sample should be taken into account while determining the total amount of 2′-deoxynucleosides. This can be determined from the HPLC trace for the enrichment (Fig. 2).

4 Notes

1. 50μL of lysis buffer containing 20 mM Tris (pH 8.1), 20 mM EDTA, 400 mM NaCl, 1% SDS (w/v), and 10μL of proteinase K (20 mg/mL) are for about 1×10^6 cells. The amounts can be adjusted for different numbers of cells.
2. The mixture should be homogeneous and clear before adding the saturated NaCl solution.
3. Better yields are achieved if the ethanol is kept at −20 °C before use, and ethanol should be placed in an ice bucket when being used.
4. Removing RNA is an important step for DNA extraction. Poorly preserved RNase A will lose its enzymatic activity. If possible, use freshly prepared aliquots of RNase A.
5. Increasing the volume of aqueous layer helps reduce the sample loss during chloroform extraction.
6. Prior to LC-MS/MS/MS analysis, acetonitrile precipitation should be performed to remove the salts in the samples. This will help increase the detection sensitivities of nucleosides and extend the lifetime of columns used.
7. The working parameters may need to be optimized on different mass spectrometers.

Acknowledgement

The author thanks for the financial support from National Institutes of Health (NIH): NIH [CA210072].

References

1. Petronis A (2010) Epigenetics as a unifying principle in the aetiology of complex traits and diseases. Nature 465:721–727
2. Jaenisch R, Bird A (2003) Epigenetic regulation of gene expression: how the genome integrates intrinsic and environmental signals. Nat Genet 33:245–254

3. Wu H, Zhang Y (2014) Reversing DNA methylation: mechanisms, genomics, and biological functions. Cell 156:45–68
4. Aruoma OI (1998) Free radicals, oxidative stress, and antioxidants in human health and disease. J Am Oil Chem Soc 75:199–212
5. Valko M, Leibfritz D, Moncol J et al (2007) Free radicals and antioxidants in normal physiological functions and human disease. Int J Biochem Cell Biol 39:44–84
6. O'Neill P (1987) The chemical basis of radiation biology. Int J Radiat Biol 52:976–976
7. Dizdaroglu M, Jaruga P (2012) Mechanisms of free radical-induced damage to DNA. Free Radic Res 46:382–419
8. Ito S, D'alessio AC, Taranova OV et al (2010) Role of Tet proteins in 5mC to 5hmC conversion, ES-cell self-renewal and inner cell mass specification. Nature 466:1129–1133
9. Tahiliani M, Koh KP, Shen Y et al (2009) Conversion of 5-methylcytosine to 5-hydroxymethylcytosine in mammalian DNA by MLL partner TET1. Science 324:930
10. Ito S, Shen L, Dai Q et al (2011) Tet proteins can convert 5-methylcytosine to 5-formylcytosine and 5-carboxylcytosine. Science 333:1300
11. He Y-F, Li B-Z, Li Z et al (2011) Tet-mediated formation of 5-carboxylcytosine and its excision by TDG in mammalian DNA. Science 333:1303
12. Cortellino S, Xu J, Sannai M et al (2011) Thymine DNA glycosylase is essential for active DNA demethylation by linked Deamination-Base excision repair. Cell 146:67–79
13. Junjie u G, Su Y, Zhong C et al (2011) Hydroxylation of 5-methylcytosine by TET1 promotes active DNA demethylation in the adult brain. Cell 145:423–434
14. Nabel CS, Jia H, Ye Y et al (2012) AID/APOBEC deaminases disfavor modified cytosines implicated in DNA demethylation. Nat Chem Biol 8:751–758
15. Liu S, Wang J, Su Y et al (2013) Quantitative assessment of Tet-induced oxidation products of 5-methylcytosine in cellular and tissue DNA. Nucleic Acids Res 41:6421–6429
16. Miller SA, Dykes DD, Polesky HF (1988) A simple salting out procedure for extracting DNA from human nucleated cells. Nucleic Acids Res 16:1215

Chapter 2

Determination of Cytosine Modifications in DNA by Chemical Labeling-Mass Spectrometry Analysis

Qing-Yun Cheng and Bi-Feng Yuan

Abstract

In active DNA demethylation, DNA cytosine methylation (5-methylcytosine, 5-mC) can be converted to 5-hydroxymethylcytosine (5-hmC), 5-formylcytosine (5-fC), and 5-carboxylcytosine (5-caC) by ten-eleven translocation (TET) proteins. These cytosine derivatives play important functions in various biological processes. 5-HmC, 5-fC, and 5-caC in genomic DNA are generally present in low abundance, thus making the quantification of these DNA modifications a challenging task. Here, we developed a method that is capable of determining all the four cytosine modifications in genomic DNA by 2-bromo-1-(4-dimethylamino-phenyl)-ethanone (BDAPE) labeling in combination with liquid chromatography-electrospray ionization tandem mass spectrometry (LC-ESI-MS/MS) analysis. This method enables the sensitive and simultaneous detection of 5-mC, 5-hmC, 5-fC, and 5-caC in genomic DNA.

Keywords 5-Methylcytosine, 5-Hydroxymethylcytosine, 5-Formylcytosine, 5-Carboxylcytosine, Chemical labeling, Mass spectrometry

1 Introduction

As one of the most important DNA epigenetic modifications, 5-methylcytosine (5-mC) plays crucial functional roles in diverse biological processes [1–4]. DNA methylation is dynamically regulated by DNA methyltransferases and demethylases [5, 6]. The active demethylation of 5-mC is involved in stepwise oxidation of 5-mC by ten-eleven translocation (TET) dioxygenases to produce 5-hydroxymethylcytosine (5-hmC), 5-formylcytosine (5-fC), and 5-carboxlxytosine (5-caC) [7–9]. Then the thymine DNA glycosylase (TDG) cleaves the *N*-glycosidic bond of 5-fC and 5-caC to form abasic sites, which are processed via DNA base excision repair (BER) pathway to restore unmodified cytosine [10–12].

In addition to be an intermediate product during active DNA demethylation, 5-hmC has been viewed as an important epigenetic mark for its regulatory function in gene expression [13–16]. Genome-wide profiling of 5-hmC revealed that 5-hmC mainly

Bi-Feng Yuan (ed.), *DNA Modification Detection Methods*, Springer Protocols Handbooks,
https://doi.org/10.1007/978-1-0716-1229-3_2,

distributed in the actively expressed gene bodies [17]. The abundance of 5-hmC has been found decreased in many kinds of cancers, indicating its involvement in tumorigenesis [18–22]. It was also reported that both 5-fC and 5-caC may be involved in DNA damage response and gene regulation [23–26].

Quantification of 5-mC and its oxidized derivatives is essential to elucidate their functional roles [27, 28]. However, the contents of 5-mC and its derivatives generally exist in low abundance in vivo, especially for 5-hmC, 5-fC, and 5-caC that are generally present in DNA in the levels of several modifications per million nucleosides [29–33]. Therefore, highly sensitive method for identification and determination of 5-mC and its oxidized derivatives is indispensable. Mass spectrometry is frequently employed for the determination of nucleic acid modifications due to its good capability on detection of trace level of compounds [34–40]. Herein, we used 2-bromo-1-(4-dimethylamino-phenyl)-ethanone (BDAPE) to simultaneously label all the four cytosine modifications (5-mC, 5-hmC, 5-fC, and 5-caC) followed by liquid chromatography-electrospray ionization tandem mass spectrometry (LC-ESI-MS/MS) analysis. The method offers the sensitive and simultaneous determination of cytosine modifications in DNA [41].

2 Materials

Water was prepared from Milli-Q apparatus (Millipore). Chromatographic-grade methanol and acetonitrile (ACN) were used. All other solvents and chemicals are of analytical grade.

2.1 Enzymes

1. S1 nuclease: 180 units/μL.
2. Venom phosphodiesterase I: 0.002 units/μL.
3. DNase I: 5 units/uL.
4. Alkaline phosphatase (CIAP): 30 units/μL.

2.2 Stock Solution of Standards

1. Thymidine (T): 500 μg/mL in water.
2. 2′-Deoxyadenosine (dA): 500 μg/mL in water.
3. 2′-Deoxyguanosine (dG): 500 μg/mL in water.
4. 2′-Deoxycytidine (dC): 500 μg/mL in water.
5. 5-Methyl-2′-deoxycytidine (5-mdC): 50 μg/mL in water.
6. 5-Hydroxymethyl-2′-deoxycytidine (5-hmdC): 10 μg/mL in water.
7. 5-Formyl-2′-deoxycytidine (5-fdC): 1 μg/mL in water.
8. 5-Carboxyl-2′-deoxycytidine (5-cadC): 1 μg/mL in water.
9. 2-Bromo-1-(4-dimethylamino-phenyl)-ethanone (BDAPE): 40 mmol/L in ACN.

10. Triethylamine (TEA): 100 mmol/L in ACN.

2.3 Chemicals and Reagents

1. Chloroform.
2. Formic acid.
3. Solvent A: formic acid in water (0.1%, v/v).
4. Solvent B: acetonitrile (ACN).
5. 10× enzymatic buffer: 3 μL (500 mM Tris-HCl, 100 mM NaCl, 10 mM $MgCl_2$, 10 mM $ZnSO_4$, pH 7.0).
6. Graphitized carbon black (Weltech Co., Ltd., Wuhan, China).
7. E.Z.N.A.® FFPE DNA Kit (Omega Bio-Tek Inc., Norcross, GA).
8. Collect formalin-fixed, paraffin-embedded (FFPE) tissue samples from colorectal carcinoma (CRC) patients in Hubei Cancer Hospital (Wuhan, China).

2.4 Equipment

1. Shimadzu LC-20 AD HPLC with a CTO-20 AC thermostated column compartment, two LC-20 AD pumps, a DGU-20A3 degasser, and a SIL-20A autosampler.
2. AB 3200 QTRAP mass spectrometer (Applied Biosystems) with an electrospray ionization source (Turbo Ionspray).
3. Hisep C18-T column (150 mm × 2.1 mm i.d., 5 μm, Weltech Co., Wuhan, China).
4. B-500 spectrophotometer (Metash Instruments Co., Shanghai, China).

3 Methods

3.1 DNA Extraction

1. Isolation of FFPE tissues DNA using E.Z.N.A.® FFPE DNA Kit according to the manufacturer's recommended procedure. Generally, the yield of isolated genomic DNA is ~3 μg per 10 mg of FFPE tissue (*see* **Note 1**).
2. Use absorbance at 260 nm and 280 nm to determine the purity and concentration of the isolated DNA by B-500 spectrophotometer (*see* **Note 2**).

3.2 Enzymatic Digestion of DNA

1. Denature 10 μg of genomic DNA (dissolved in 21 μL of H_2O) by heating at 95 °C for 5 min and then chilling on ice for 2 min.
2. Add 3 μL of 10× enzymatic buffer, 2 μL of S1 nuclease, 2 μL of venom phosphodiesterase I, 1 μL of alkaline phosphatase, and 1 μL of DNase I. Gently flick the tube with fingers, shortly centrifuge, and then incubate the mixture (30 μL) at 37 °C for 4 h (*see* **Note 3**).

3. Add 270 μL of H_2O and 300 μL of chloroform to the tube (*see* **Note 4**). Vigorous vortex of the mixture for 3 min and then centrifugation at 12000*g* for 10 min. Collect the supernatant to a clean tube and add 300 μL of chloroform for next round of extraction. Repeat the extraction for three times (*see* **Note 5**).

3.3 Remove Salts in Digestion Mixture with Graphitized Carbon Black

1. Pack the SPE (solid phase extraction) cartridge with 50 mg sorbent of graphitized carbon black (Weltech Co., Ltd., Wuhan, China).
2. Condition the cartridge successively with 2 mL of CH_2Cl_2-MeOH (4:1, v/v), 2 mL of H_2O. Keep the sorbent in H_2O before sampling of DNA.
3. Pass the digested mixture (dissolved in H_2O) through the SPE cartridge. Digested nucleosides will retain on the column (*see* **Note 6**).
4. Wash the cartridge by adding 2 mL of H_2O to remove salts. Dry the cartridge under vacuum.
5. Digested nucleosides are eluted by 100 μL of MeOH and 1 mL of CH_2Cl_2-MeOH (4:1, v/v) in sequence. Combine the desorption solution.
6. Dry the elution with nitrogen gas at 37 °C.

3.4 Chemical Labeling

The chemical reaction of BDAPE with 5-mdC, 5-hmdC, 5-fdC, and 5-cadC is shown in Fig. 1.

1. The dried nucleosides are labeled by 4 mM of BDAPE in 200 μL of ACN with TEA (4 mM) as the catalyst. The reaction is carried out under 60 °C for 6 h with shaking at 1500 rpm.
2. Dry the labeled products with nitrogen gas at 37 °C. Redissolve the mixture in 100 μL water followed by subsequent LC-E-SI-MS/MS analysis.

3.5 Determination of the Derivatives of 5-mdC, 5-hmdC, 5-fdC, and 5-cadC by LC-ESI-MS/MS

1. Analysis of the chemically labeled derivatives of 5-mdC, 5-hmdC, 5-fdC, and 5-cadC is performed on a Shimadzu LC-20 AD HPLC system coupled with an AB 3200 QTRAP mass spectrometer.
2. The HPLC separation of the four derivatives is conducted on a Hisep C18-T column with a flow rate of 0.2 mL/min at 35 °C.
3. Mobile phases are consisted of solvent A and solvent B. A gradient of 5–65% B for 40 min is used (*see* **Note 7**).
4. Positive mass spectrometry detection mode is used.
5. Monitor the analytes under multiple reaction monitoring (MRM) mode. Mass transitions (precursor ions → product ions) are as follows: 5-mdC derivatives (385.2 → 269.1), 5-hmdC derivatives (401.2 → 285.1 and 401.2 → 267.1),

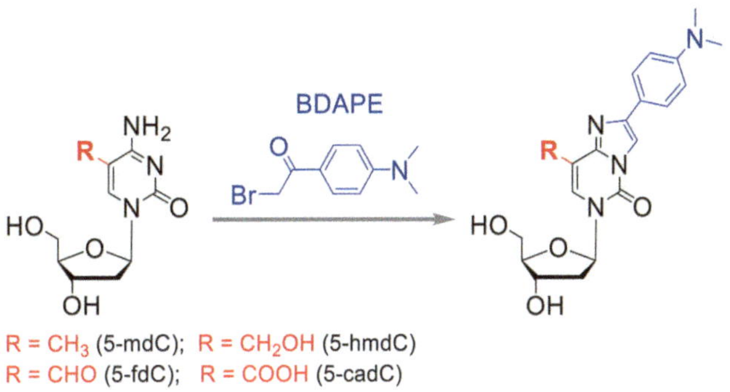

Fig. 1 Chemical labeling of 5-mdC, 5-hmdC, 5-fdC, and 5-cadC by BDAPE. (Reproduced from ref. [41] with permission from American Chemical Society)

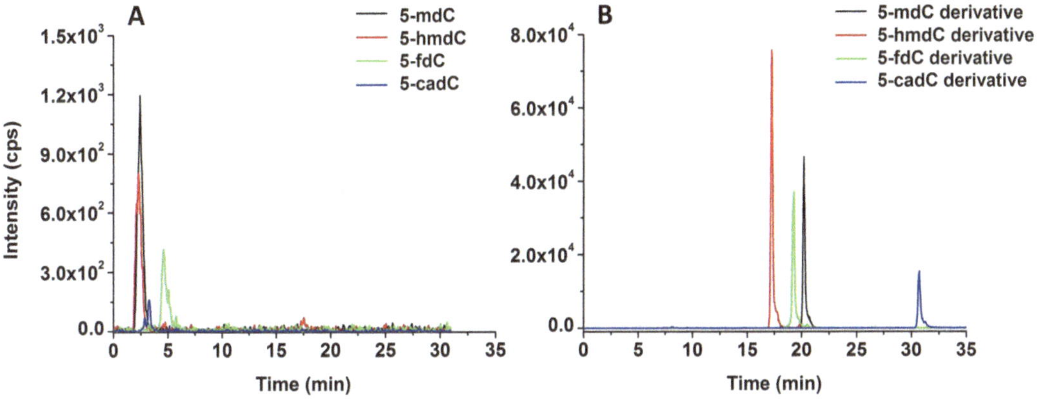

Fig. 2 MRM extracted-ion chromatograms of 5-mdC, 5-hmdC, 5-fdC, and 5-cadC before (**a**) and after (**b**) BDAPE labeling under optimized conditions. (Reproduced from ref. [41] with permission from American Chemical Society)

5-fdC derivatives (399.2 → 283.1), 5-cadC derivatives (576.2 → 460.2 and 576.2 → 281.1), T (243.1 → 127.1), dC (228.1 → 112.1), dA (252.1 → 136.1), dG (268.1 → 152.1), 5-mdC (242.1 → 126.1), 5-hmdC (258.1 → 142.1), 5-fdC (256.1 → 140.0), and 5-cadC (272.1 → 156.0).

6. The MRM parameters of all nucleosides are optimized to achieve maximal detection sensitivity.

7. AB SCIEX Analyst 1.5 Software is employed for data acquisition and processing.

8. Figure 2a shows the extracted-ion chromatogram of 5-mdC, 5-hmdC, 5-fdC, and 5-cadC. Without BDAPE labeling, they are eluted from the C18 chromatographic column within 6 min. 5-mdC, 5-hmdC, and 5-cadC cannot be well separated under optimized conditions. Figure 2b shows the extracted-ion chromatogram of 5-mdC, 5-hmdC, 5-fdC, and 5-cadC

Table 1
LODs of 5-mdC, 5-hmdC, 5-fdC, and 5-cadC with and without BDAPE labeling. (Reproduced from ref. [41] with permission from American Chemical Society)

	LOD (fmol)			
	5-mdC	5-hmdC	5-fdC	5-cadC
Without BDAPE labeling	3.5	5.6	9.8	28.4
After BDAPE labeling	0.10	0.06	0.11	0.23
Increased folds of the detection sensitivity	35	93	89	123

derivatives. After BDAPE labeling, these derivatives have longer retention times due to the introduction of hydrophobic group from BDAPE. Moreover, significant improvement in chromatographic resolution of these derivatives is achieved.

9. The limit of detection (LOD), which is defined as the amounts of the analyte when the signal/noise ratios is 3, is employed to evaluate the detection sensitivity. After BDAPE labeling, the detection sensitivities of 5-mdC, 5-hmdC, 5-fdC, and 5-cadC dramatically increased by 35, 93, 89, and 123 folds, respectively (Table 1).

3.6 Construction of Calibration Curve

1. Calibration curves are constructed by plotting the mean peak area ratio of 5-mdC/10^2 dG, 5-hmdC/10^5 dG, 5-fdC/10^7 dG, or 5-cadC/10^7 dG versus the mean molar ratios of 5-mdC/10^2 dG, 5-hmdC/10^5 dG, 5-fdC/10^7 dG, or 5-cadC/10^7 dG from triplicate measurements. The calibration curve data are shown in Table 2.

3.7 Simultaneous Determination of 5-mdC, 5-hmdC, 5-fdC, and 5-cadC in Genomic DNA of Human CRC Tissues

1. The process of the developed method consists of DNA extraction, enzymatic digestion, removing of salts in DNA with graphitized carbon black-SPE, BDAPE labeling followed by LC-ESI-MS/ MS analysis.
2. With the developed method, 5-mdC, 5-hmdC, 5-fdC, and 5-cadC have been simultaneously detected and quantified in 24 tumor tissue samples derived from 24 CRC patients. The mean contents of 5-mdC, 5-hmdC, 5-fdC, and 5-cadC in human CRC tumor tissues are $3.5 \pm 0.6/10^2$ dG, $27.0 \pm 6.4/10^5$ dG, $9.2 \pm 3.4/10^7$ dG, and $0.9 \pm 0.3/10^7$ dG, respectively.

4 Notes

1. This method can be used for isolating DNA from other types of tissues. The yield varies according to the type of tissue used.

Table 2
Linearities of 5-mdC, 5-hmdC, 5-fdC, and 5-cadC by BDAPE labeling combined with LC-ESI-MS/MS analysis. (Reproduced from ref. [41] with permission from American Chemical Society)

		Calibration curve data		
Compound molar ratio	**Linear range**	**Slope**	**Intercept**	***R* value**
5-mdC/10^2 dG	0.1–10	3.7624	−0.9065	0.9966
5-hmdC/10^5 dG	2–500	0.0028	−0.0104	0.9978
5-fdC/10^7 dG	0.5–100	0.00006	0.00007	0.9983
5-cadC/10^7 dG	0.5–100	0.00002	−0.000007	0.9962

2. The values of OD260/280 ranging from 1.8 to 2.0 indicate the good quality of isolated DNA.
3. Do not mix the enzymes by vigorous vortex because the activity of enzymes may decrease by vigorous vortex.
4. The enzymes in the digested DNA can be precipitated by extraction with equal volume of chloroform. If the amount of isolated DNA is limited, more water can be added into the reaction mixture to reduce the DNA loss.
5. If there is any protein in the aqueous layer after chloroform extraction for three times, it is necessary to repeat the extraction for additional times.
6. With the help of vacuum pump, the speed of liquid flowing through the SPE cartridge should be kept at ~100 μL/min.
7. Generally, the nucleosides can form the metal ion adducts, such as K^+ and Na^+ adducts. Addition of formic acid in mobile phases can reduce the occurrence of these adducts.

Acknowledgements

The work is supported by the National Natural Science Foundation of China (22074110, 21672166, 21721005).

References

1. Luo C, Hajkova P, Ecker JR (2018) Dynamic DNA methylation: in the right place at the right time. Science 361:1336–1340
2. Liu T, Ma CJ, Yuan BF, Feng YQ (2018) Modificaomics: deciphering the functions of biomolecule modifications. Sci China Chem 61:381–392
3. Dor Y, Cedar H (2018) Principles of DNA methylation and their implications for biology and medicine. Lancet 392:777–786
4. Allis CD, Jenuwein T (2016) The molecular hallmarks of epigenetic control. Nat Rev Genet 17:487–500
5. Shen L, Song CX, He C, Zhang Y (2014) Mechanism and function of oxidative reversal

of DNA and RNA methylation. Annu Rev Biochem 83:585–614
6. Chen Z, Zhang Y (2020) Role of mammalian DNA methyltransferases in development. Annu Rev Biochem 89:135–158
7. Kohli RM, Zhang Y (2013) TET enzymes, TDG and the dynamics of DNA demethylation. Nature 502:472–479
8. Bilyard MK, Becker S, Balasubramanian S (2020) Natural, modified DNA bases. Curr Opin Chem Biol 57:1–7
9. Carell T, Kurz MQ, Muller M, Rossa M, Spada F (2018) Non-canonical bases in the genome: the regulatory information layer in DNA. Angew Chem Int Ed Engl 57:4296–4312
10. Feng Y, Xie NB, Tao WB, Ding JH, You XJ, Ma CJ, Zhang X, Yi C, Zhou X, Yuan BF, Feng YQ (2020) Transformation of 5-carboxylcytosine to cytosine through C–C bond cleavage in human cells constitutes a novel pathway for DNA demethylation. CCS Chem 2:994–1008
11. Wu X, Zhang Y (2017) TET-mediated active DNA demethylation: mechanism, function and beyond. Nat Rev Genet 18:517–534
12. Lu X, Zhao BS, He C (2015) TET family proteins: oxidation activity, interacting molecules, and functions in diseases. Chem Rev 115:2225–2239
13. Bachman M, Uribe-Lewis S, Yang X, Williams M, Murrell A, Balasubramanian S (2014) 5-hydroxymethylcytosine is a predominantly stable DNA modification. Nat Chem 6:1049–1055
14. Tang Y, Chu JM, Huang W, Xiong J, Xing XW, Zhou X, Feng YQ, Yuan BF (2013) Hydrophilic material for the selective enrichment of 5-hydroxymethylcytosine and its liquid chromatography-tandem mass spectrometry detection. Anal Chem 85:6129–6135
15. Tang Y, Xiong J, Jiang HP, Zheng SJ, Feng YQ, Yuan BF (2014) Determination of oxidation products of 5-methylcytosine in plants by chemical derivatization coupled with liquid chromatography/tandem mass spectrometry analysis. Anal Chem 86:7764–7772
16. Chen K, Zhao BS, He C (2016) Nucleic acid modifications in regulation of gene expression. Cell Chem Biol 23:74–85
17. Han D, Lu X, Shih AH, Nie J, You Q, Xu MM, Melnick AM, Levine RL, He C (2016) A highly sensitive and robust method for genome-wide 5hmC profiling of rare cell populations. Mol Cell 63:711–719
18. Storebjerg TM, Strand SH, Hoyer S, Lynnerup AS, Borre M, Orntoft TF, Sorensen KD (2018) Dysregulation and prognostic potential of 5-methylcytosine (5mC), 5-hydroxymethylcytosine (5hmC), 5-formylcytosine (5fC), and 5-carboxylcytosine (5caC) levels in prostate cancer. Clin Epigenetics 10:105
19. Pfeifer GP, Xiong W, Hahn MA, Jin SG (2014) The role of 5-hydroxymethylcytosine in human cancer. Cell Tissue Res 356:631–641
20. Chen K, Zhang J, Guo Z, Ma Q, Xu Z, Zhou Y, Xu Z, Li Z, Liu Y, Ye X, Li X, Yuan B, Ke Y, He C, Zhou L, Liu J, Ci W (2016) Loss of 5-hydroxymethylcytosine is linked to gene body hypermethylation in kidney cancer. Cell Res 26:103–118
21. Chen ML, Shen F, Huang W, Qi JH, Wang Y, Feng YQ, Liu SM, Yuan BF (2013) Quantification of 5-methylcytosine and 5-hydroxymethylcytosine in genomic DNA from hepatocellular carcinoma tissues by capillary hydrophilic-interaction liquid chromatography/quadrupole TOF mass spectrometry. Clin Chem 59:824–832
22. Kroeze LI, van der Reijden BA, Jansen JH (2015) 5-hydroxymethylcytosine: an epigenetic mark frequently deregulated in cancer. Biochim Biophys Acta 1855:144–154
23. Song CX, Szulwach KE, Dai Q, Fu Y, Mao SQ, Lin L, Street C, Li Y, Poidevin M, Wu H, Gao J, Liu P, Li L, Xu GL, Jin P, He C (2013) Genome-wide profiling of 5-formylcytosine reveals its roles in epigenetic priming. Cell 153:678–691
24. Lu X, Han D, Zhao BS, Song CX, Zhang LS, Dore LC, He C (2015) Base-resolution maps of 5-formylcytosine and 5-carboxylcytosine reveal genome-wide DNA demethylation dynamics. Cell Res 25:386–389
25. Zhu C, Gao Y, Guo H, Xia B, Song J, Wu X, Zeng H, Kee K, Tang F, Yi C (2017) Single-cell 5-formylcytosine landscapes of mammalian early embryos and ESCs at Single-Base resolution. Cell Stem Cell 20:720–731.e725
26. Zhang Y, Zhou C (2019) Formation and biological consequences of 5-formylcytosine in genomic DNA. DNA Repair (Amst) 81:102649
27. Booth MJ, Raiber EA, Balasubramanian S (2015) Chemical methods for decoding cytosine modifications in DNA. Chem Rev 115:2240–2254
28. Hofer A, Liu ZJ, Balasubramanian S (2019) Detection, structure and function of modified DNA bases. J Am Chem Soc 141:6420–6429
29. Yuan BF, Feng YQ (2014) Recent advances in the analysis of 5-methylcytosine and its oxidation products. Trend Anal Chem 54:24–35
30. Xiong J, Jiang HP, Peng CY, Deng QY, Lan MD, Zeng H, Zheng F, Feng YQ, Yuan BF

(2015) DNA hydroxymethylation age of human blood determined by capillary hydrophilic-interaction liquid chromatography/mass spectrometry. Clin Epigenetics 7:72

31. Xiong J, Yuan BF, Feng YQ (2019) Mass spectrometry for investigating the effects of toxic metals on nucleic acid modifications. Chem Res Toxicol 32:808–819
32. Yuan BF (2020) Assessment of DNA epigenetic modifications. Chem Res Toxicol 33:695–708
33. Lan MD, Yuan BF, Feng YQ (2019) Deciphering nucleic acid modifications by chemical derivatization-mass spectrometry analysis. Chin Chem Lett 30:1–6
34. Huang W, Qi CB, Lv SW, Xie M, Feng YQ, Huang WH, Yuan BF (2016) Determination of DNA and RNA methylation in circulating tumor cells by mass spectrometry. Anal Chem 88:1378–1384
35. Chen B, Yuan BF, Feng YQ (2019) Analytical methods for deciphering RNA modifications. Anal Chem 91:743–756
36. Liu FL, Qi CB, Cheng QY, Ding JH, Yuan BF, Feng YQ (2020) Diazo reagent labeling with mass spectrometry analysis for sensitive determination of ribonucleotides in living organisms. Anal Chem 92:2301–2309
37. Cheng QY, Xiong J, Wang F, Yuan BF, Feng YQ (2018) Chiral derivatization coupled with liquid chromatography/mass spectrometry for determining ketone metabolites of hydroxybutyrate enantiomers. Chin Chem Lett 29:115–118
38. Qi CB, Ding JH, Yuan BF, Feng YQ (2019) Analytical methods for locating modifications in nucleic acids. Chin Chem Lett 30:1618–1626
39. Qi CB, Jiang HP, Xiong J, Yuan BF, Feng YQ (2019) On-line trapping/capillary hydrophilic-interaction liquid chromatography/mass spectrometry for sensitive determination of RNA modifications from human blood. Chin Chem Lett 30:553–557
40. You XJ, Liu T, Ma CJ, Qi CB, Tong Y, Zhao X, Yuan BF, Feng YQ (2019) Determination of RNA hydroxylmethylation in mammals by mass spectrometry analysis. Anal Chem 91:10477–10483
41. Tang Y, Zheng SJ, Qi CB, Feng YQ, Yuan BF (2015) Sensitive and simultaneous determination of 5-methylcytosine and its oxidation products in genomic DNA by chemical derivatization coupled with liquid chromatography-tandem mass spectrometry analysis. Anal Chem 87:3445–3452

Chapter 3

Analysis of 5-Methylcytosine and 5-Hydroxymethylcytosine in Genomic DNA by Capillary Electrophoresis-Mass Spectrometry

Fang Yuan, Yu-Fang Ma, Ying-Lin Zhou, and Xin-Xiang Zhang

Abstract

5-Methylcytosine (5mC) and 5-hydroxymethylcytosine (5hmC) are important epigenetic biomarkers and are demonstrated to be promising in early cancer diagnosis. The investigation on the global levels and distribution of these two cytosine modifications can reveal their biology functions and help develop diagnostic methods based on these biomarkers. However, compared with a variety of sequencing methods which aim at investigating the distribution of cytosine modifications, HPLC-MS/MS is a powerful technique to quantify the global levels of cytosine modifications. The newly developed sheathless CESI-MS/MS was introduced in this chapter. The method is a sensitive and simple technique to evaluate the global levels of 5mC and 5hmC in different kinds of DNA samples. The protocols for the sample preparation, instrumentations, and data analysis were proposed. With CESI-MS/MS, the limits of detection (LODs) of cytosine modifications can reach attomole level. Meanwhile, CESI-MS/MS can quantify the 5mC/5hmC levels in practical DNA sample less than 10 ng. Considering the ultrahigh sensitivity and the low DNA consumption, CESI-MS/MS can either be used in the preliminary study before sequencing in a research setting or as a final detection technique for the clinical diagnosis.

Keywords Genomic DNA, 5-Methylcytosine, 5-Hydroxymethylcytosine, CESI-MS/MS

1 Introduction

5-Methylcytosine (5mC) is one of the best characterized epigenetic modifications associating with the control of gene expression [1, 2], determination of cell development [3], suppression of transposable elements [4], and disease pathogenesis [5]. Besides 5mC, 5-hydroxymethylcytosine (5hmC) is also identified to be a stable epigenetic mark involved in gene regulation and cellular development [6–8]. Recently, both 5mC and 5hmC are proved to be excellent cancer biomarkers since different 5mC/5hmC signatures are observed in many cancer types compared with the healthy controls [9–11]. Therefore, the global levels and distribution of

Bi-Feng Yuan (ed.), *DNA Modification Detection Methods*, Springer Protocols Handbooks,
https://doi.org/10.1007/978-1-0716-1229-3_3,

these cytosine modifications in genomic DNA (gDNA) are important data required in both research setting and clinical diagnosis.

High performance liquid chromatography coupled with electrospray ionization tandem mass spectrometry (HPLC-MS/MS) is an important and wildly technology used in epigenetics, and also considered as the gold standard of accurate quantification of the overall level of DNA modifications [12–14]. With the use of the multiple reaction monitoring (MRM) mode of the tandem mass spectrometry (MS/MS), the detection sensitivities for cytosine modifications can be improved [12, 15, 16]. Moreover, a lot of analytical methods aiming at improving the ionization efficiency of these nucleosides and the detecting sensitivity of these targets based on HPLC-MS/MS have been developed [17–21].

Sheathless capillary electrophoresis-electrospray ionization mass spectrometry (CESI-MS) is a newly developed analytical technique which has been successfully applied in the area of proteomics and metabolomics [22–24]. The sheathless interface coupled with a porous tip sprayer technique enhances the sensitivity of CESI-MS to the sub-nanomolar level for many organic compounds, much lower than that obtained by regular HPLC-MS/MS [25, 26]. Besides high sensitivity, the CE system also offers a lot of inherent advantages to the combination of CESI-MS. It has high separation efficiency, diverse separation modes for different types of samples and low sample consumption. These properties make CE an excellent separation method for polar and charged small molecules such as the cytosine modifications we are interested in.

To sensitively detect cytosine modifications in gDNA samples and accurately quantify their global levels, we coupled CESI system with tandem mass spectrometry, and developed an ultrasensitive CESI-MS/MS method. With CESI-MS/MS, both 5mC and 5hmC can be detected in less than a few nanograms of DNA samples (Fig. 1).

2 Materials

2.1 Reagents and Solutions

1. Acetic acid (CH_3COOH, 99.9985%, metal basis) was purchased from Alfa Aesar (Ward Hill, MA, USA). 10% acetic acid (v/v, pH 2.2) was prepared by mixed 1 mL acetic acid and 9 mL H_2O together. Prepare freshly before use (*see* **Note 1**).
2. Methanol (LC-MS Grade) was purchased from J.T. Baker. Store at 25 °C.
3. Hydrochloric acid were purchased from Xilong Scientific (Guangdong, China). 0.1 M hydrochloric acid was prepared by mixing 0.1 mL 12 M hydrochloric acid with 11.9 mL H_2O. Store at 25 °C.

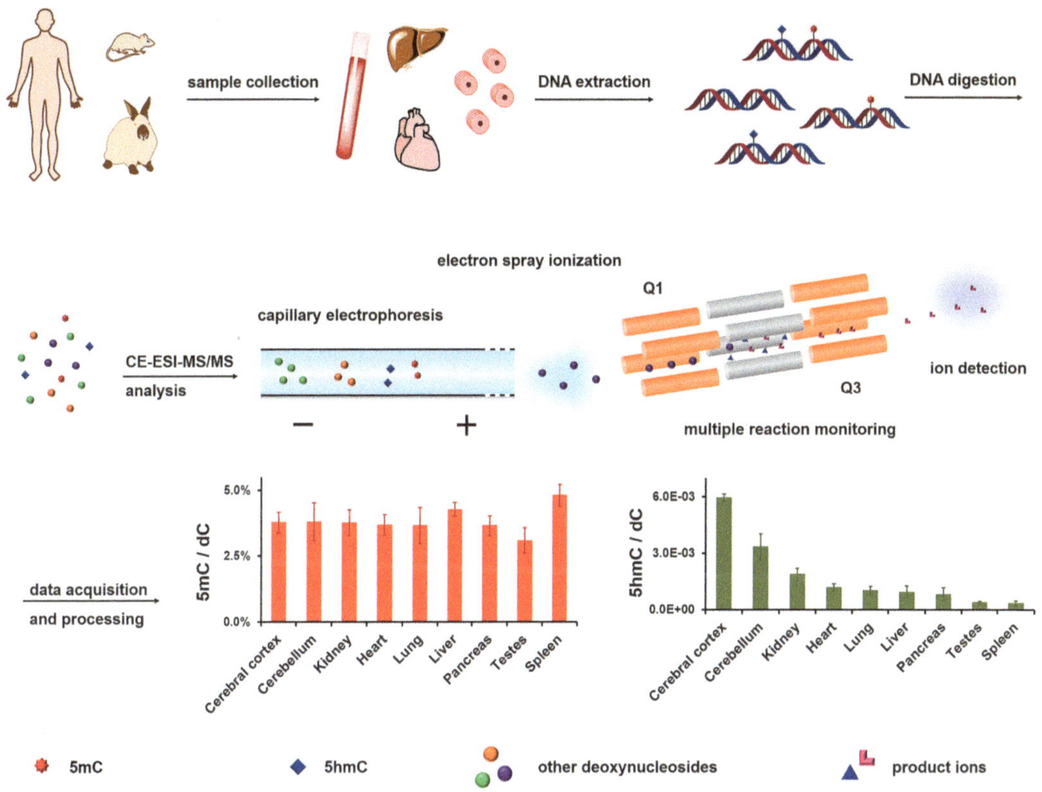

Fig. 1 Procedure for the analysis of 5mC and 5hmC in genomic DNA using CE-ESI-MS/MS. (Reproduced from ref. [27] with permission from The Royal Society of Chemistry and from ref. [28])

4. Sodium hydroxide was purchased from Beijing Tong Guang Fine Chemicals Company (Beijing, China). 0.1 M sodium hydroxide was prepared by adding 10 mL H_2O into 0.04 g sodium hydroxide. Store at 25 °C.
5. Water used in this work was purified using a Milli-Q water purification system. Store at 25 °C.

2.2 Equipment

1. Nanodrop 2000 spectrophotometer (Thermo Scientific, US).
2. CESI 8000 Plus ESI-MS (SCIEX, US).
3. 5500 Triple Quadrupole Mass Spectrometer (SCIEX, US).
4. Milli-Q water purification system (Millipore, US).
5. Centrifuge tubes, 1.5 and 15 mL (Coring, US).
6. OptiMS CESI cartridge (SCIEX, US).
7. NanoVial (SCIEX, US).

2.3 Software

1. Analyst software (SCIEX, US).
2. MultiQuant 3.0.2 Software (SCIEX, US).

2.4 Enzymes and Kit

1. TaKaRa MiniBEST Universal Genomic DNA Extraction Kit Ver.5.0 (Takara, Biotechnology Co., Ltd. Dalian, China).
2. Nucleoside Digestion Mix (New England BioLabs).

3 Methods

3.1 Extraction and Purification of DNA

1. Collect 2–25 mg animal tissues, 50–100μL blood samples, or 1.0×10^5–1.0×10^7 cells.
2. Extract genomic DNA of the tissue, blood or cell samples using the TaKaRa MiniBEST Universal Genomic DNA Extraction Kit Ver.5.0 (Takara, Biotechnology Co., Ltd. Dalian, China) according to the manufacturer's instructions.
3. Analyze the concentration and quality of the extracted gDNA samples using Nanodrop 2000 spectrophotometer. Typically, 1–10μg gDNA can be obtained using the above sample and procedure. The A260/A280 value of the extracted DNA samples should be in the range of 1.8–2.0.
4. Store the DNA at −80 °C or proceed to DNA digestion step.

3.2 DNA Digestion

1. Take 100 ng–1μg gDNA into a 1.5 mL centrifuge tube for Nucleoside Digestion Mix (New England BioLabs) digestion.
2. Add 1μL of Nucleoside Digestion Mix, 1μL of 10 × Nucleoside Digestion Mix Reaction buffer and certain volume of H_2O to the tube to a final reaction volume of 10μL.
3. Incubate at 37 °C for 1 h.
4. Centrifuge at 10000 ×*g* for 2 min at room temperature, then transfer the digested solution into a SCIEX nanoVial for CESI-MS analysis.

3.3 CESI-MS/MS Analysis

After DNA extraction and enzymatic digestion, the obtained deoxynucleosides are analyzed by CESI-MS/MS (*see* **Note 2**). CESI-MS/MS analysis is performed with a CESI 8000 plus capillary electrophoresis system coupled with a Triple Quad 5500 mass spectrometry through the NanoSpray III source. The system is controlled by Analyst software. For CESI 8000 plus capillary electrophoresis system, a cartridge equipped with two bare fused-silica capillaries, namely separation capillary and conductive liquid capillary, is used for the separation. The separation capillary is 91-cm long with an internal diameter of 30μm and an outside diameter of 150μm, and it is etched with a porous tip. The porous tip is then inserted into the nanospray interface (*see* **Notes 3, 4**, and **5**).

1. At the very first, the separation capillary needs to be flushed and activated for better performance. It is sequentially flushed with methanol, H_2O, 0.1 M sodium hydroxide, 0.1 M hydrochloric acid and H_2O for 10 min at 100 psi each.

2. To fit the separation conditions, the separation capillary and conductive liquid capillary are rinsed with 10% acetic acid (v/v, pH 2.2), which is also used as the background electrolyte (BGE), at 100 psi for 10 min and 5 min, respectively.
3. Store the samples at 4 °C until analysis.
4. The detailed parameters of the CE separation system are described as follows:

   Set the temperature of the sample plate at 4 °C.

   Use hydrodynamic injection with a pressure of 2.5 psi for 30 s.

   Set the temperature of the cartridge at 25 °C.

   Set the separation voltage at +25 kV and the current is between 3.0 and 3.2μA.
5. For mass spectrometry, ESI is performed in positive ionization mode. The ionspray voltage is set as +1750 eV for the best nanospray stability and efficiency. The values for curtain gas, gas 1, gas 2, temperature, declustering potential, entrance potential, and collision cell exit potential are optimized to be 5 psi, 0 psi, 0 psi, 0 °C, 80 V, 10 V, 13 V, respectively. Multiple reaction monitoring (MRM) mode is used for the detection and quantification of the deoxynucleosides. For 5mC, the precursor ion (Q1), product ion (Q3), and collision energy are 242.1, 126.1, and 16.0; for 5hmC, they are 258.1, 142.1, and 15.0, respectively.
6. For the quantification of the 5mC and 5hmC in DNA samples, calibration curves are constructed firstly using mixture solution of 5mC and 5hmC standards in different concentrations from 80 pM to 600 nM (Table 1). For each concentration, an average value of three independent prepared samples is determined. MultiQuant 3.0.2 Software is used for the data processing. The 5mC/dC and 5hmC/(dC + 5mC) results of each sample are then calculated.
7. Genomic DNA of several tissues samples of three rats are extracted, digested, and analyzed using CESI-MS/MS. The results are shown in Table 2.

Table 1
CESI-MS/MS calibration curves and linear range of 5mC and 5hmC

Substance	Calibration curve (c/nM)	R^2	Linear range (nM)
dC	Area = 2.18×10^4 c − 1.22×10^4	0.999	25.0–600
5mC	Area = 4.06×10^4 c − 7.18×10^4	0.993	2.0–60
5hmC	Area = 2.76×10^4 c + 8.05×10^2	0.999	0.080–3.0

Table 2
CESI-MS/MS quantification results of 5mC and 5hmC of tissue gDNA samples

		dC/nM	5mC/dC (%)	5hmC/(dC + 5mC) (%)
Heart	A1	332.15	4.11	0.139
	B1	38.56	3.63	0.100
	C1	107.46	3.31	0.117
Liver	A2	365.41	3.99	0.095
	B2	95.96	4.32	0.060
	C2	68.64	4.52	0.125
Spleen	A3	603.91	4.96	0.050
	B3	298.08	4.35	0.023
	C3	368.26	5.14	0.031
Kidney	A4	169.52	4.34	0.187
	B4	74.51	3.48	0.220
	C4	49.73	3.48	0.155
Brain cortex	A5	147.84	4.17	0.597
	B5	55.36	3.76	0.575
	C5	25.15	3.38	0.615
Cerebellum	A6	447.8	3.00	0.284
	B6	128.87	4.42	0.305
	C6	60.67	3.97	0.412
Pancreas	A7	463.76	3.85	0.075
	B7	87.51	3.90	0.121
	C7	152.63	3.22	0.051

4 Notes

1. For good reproducibility, 10% acetic acid (v/v, pH 2.2) need to be prepared freshly before use.
2. Ensure there isn't any air bubble left in the vial after adding the digested solution into the SCIEX nanoVial for CESI-MS analysis.
3. During the analysis by CESI-MS/MS, pay attention to the current and ensure that it is stable.

4. For the best detection stability and sensitivity, MS parameters should be optimized carefully. We use the standard solution at 10μM (in 10% acetic acid) for the optimization.
5. After the whole analysis, flush the capillaries with methanol and pump out the methanol using negative pressure.

Acknowledgements

This work was supported by the National Natural Science 443Foundation of China (Grant Nos. 21675004, 21575005, and 44421775006).

References

1. Bird A (2002) DNA methylation patterns and epigenetic memory. Genes Dev 16(1):6–21
2. Jaenisch R, Bird A (2003) Epigenetic regulation of gene expression: how the genome integrates intrinsic and environmental signals. Nat Genet 33:245–254
3. Smith ZD, Meissner A (2013) DNA methylation: roles in mammalian development. Nat Rev Genet 14(3):204–220
4. Rottach A, Leonhardt H, Spada F (2009) DNA methylation-mediated epigenetic control. J Cell Biochem 108(1):43–51
5. Robertson KD (2005) DNA methylation and human disease. Nat Rev Genet 6(8):597–610
6. Kriaucionis S, Heintz N (2009) The nuclear DNA base 5-hydroxymethylcytosine is present in Purkinje neurons and the brain. Science 324 (5929):929–930
7. Branco MR, Ficz G, Reik W (2011) Uncovering the role of 5-hydroxymethylcytosine in the epigenome. Nat Rev Genet 13(1):7–13
8. Kriukiene E, Liutkeviciute Z, Klimasauskas S (2012) 5-hydroxymethylcytosine—the elusive epigenetic mark in mammalian DNA. Chem Soc Rev 41(21):6916–6930
9. Chan KC, Jiang P, Chan CW, Sun K, Wong J, Hui EP, Chan SL, Chan WC, Hui DS, Ng SS, Chan HL, Wong CS, Ma BB, Chan AT, Lai PB, Sun H, Chiu RW, Lo YM (2013) Noninvasive detection of cancer-associated genome-wide hypomethylation and copy number aberrations by plasma DNA bisulfite sequencing. Proc Natl Acad Sci U S A 110(47):18761–18768
10. Wen L, Li J, Guo H, Liu X, Zheng S, Zhang D, Zhu W, Qu J, Guo L, Du D, Jin X, Zhang Y, Gao Y, Shen J, Ge H, Tang F, Huang Y, Peng J (2015) Genome-scale detection of hypermethylated CpG islands in circulating cell-free DNA of hepatocellular carcinoma patients. Cell Res 25(11):1250–1264
11. Wan JCM, Massie C, Garcia-Corbacho J, Mouliere F, Brenton JD, Caldas C, Pacey S, Baird R, Rosenfeld N (2017) Liquid biopsies come of age: towards implementation of circulating tumour DNA. Nat Rev Cancer 17(4):223–238
12. Liu S, Wang J, Su Y, Guerrero C, Zeng Y, Mitra D, Brooks PJ, Fisher DE, Song H, Wang Y (2013) Quantitative assessment of Tet-induced oxidation products of 5-methylcytosine in cellular and tissue DNA. Nucleic Acids Res 41 (13):6421–6429
13. Yuan B-F, Feng Y-Q (2014) Recent advances in the analysis of 5-methylcytosine and its oxidation products. Trends Anal Chem 54:24–35
14. Hofer A, Liu ZJ, Balasubramanian S (2019) Detection, structure and function of modified DNA bases. J Am Chem Soc 141 (16):6420–6429
15. Singh R, Farmer PB (2006) Liquid chromatography-electrospray ionization-mass spectrometry: the future of DNA adduct detection. Carcinogenesis 27(2):178–196
16. Ito S, Shen L, Dai Q, Wu SC, Collins LB, Swenberg JA, He C, Zhang Y (2011) Tet proteins can convert 5-methylcytosine to 5-formylcytosine and 5-carboxylcytosine. Science 333(6047):1300–1303
17. Zhang JJ, Zhang L, Zhou K, Ye X, Liu C, Zhang L, Kang J, Cai C (2011) Analysis of global DNA methylation by hydrophilic interaction ultra high-pressure liquid chromatography tandem mass spectrometry. Anal Biochem 413(2):164–170
18. Tang Y, Xiong J, Jiang HP, Zheng SJ, Feng YQ, Yuan BF (2014) Determination of oxidation products of 5-methylcytosine in plants by chemical derivatization coupled with liquid

chromatography/tandem mass spectrometry analysis. Anal Chem 86(15):7764–7772

19. Tang Y, Zheng SJ, Qi CB, Feng YQ, Yuan BF (2015) Sensitive and simultaneous determination of 5-methylcytosine and its oxidation products in genomic DNA by chemical derivatization coupled with liquid chromatography-tandem mass spectrometry analysis. Anal Chem 87(6):3445–3452
20. Yin R, Mo J, Lu M, Wang H (2015) Detection of human urinary 5-hydroxymethylcytosine by stable isotope dilution HPLC-MS/MS analysis. Anal Chem 87(3):1846–1852
21. Shahal T, Koren O, Shefer G, Stern N, Ebenstein Y (2018) Hypersensitive quantification of global 5-hydroxymethylcytosine by chemoenzymatic tagging. Anal Chim Acta 1038:87–96
22. Sarg B, Faserl K, Kremser L, Halfinger B, Sebastiano R, Lindner HH (2013) Comparing and combining capillary electrophoresis electrospray ionization mass spectrometry and nano-liquid chromatography electrospray ionization mass spectrometry for the characterization of post-translationally modified histones. Mol Cell Proteomics 12(9):2640–2656
23. Zhu G, Sun L, Yan X, Dovichi NJ (2013) Single-shot proteomics using capillary zone electrophoresis-electrospray ionization-tandem mass spectrometry with production of more than 1250 Escherichia coli peptide identifications in a 50 min separation. Anal Chem 85(5):2569–2573
24. Ibrahim M, Gahoual R, Enkler L, Becker HD, Chicher J, Hammann P, Francois YN, Kuhn L, Leize-Wagner E (2016) Improvement of mitochondria extract from Saccharomyces cerevisiae characterization in shotgun proteomics using sheathless capillary electrophoresis coupled to tandem mass spectrometry. J Chromatogr Sci 54(4):653–663
25. Ramautar R, Busnel JM, Deelder AM, Mayboroda OA (2012) Enhancing the coverage of the urinary metabolome by sheathless capillary electrophoresis-mass spectrometry. Anal Chem 84(2):885–892
26. Wang Y, Fonslow BR, Wong CC, Nakorchevsky A, Yates JR 3rd (2012) Improving the comprehensiveness and sensitivity of sheathless capillary electrophoresis-tandem mass spectrometry for proteomic analysis. Anal Chem 84(20):8505–8513
27. Yuan F, Zhang X-H, Nie J, Chen H-X, Zhou Y-L, Zhang X-X (2016) Ultrasensitive determination of 5-methylcytosine and 5-hydroxymethylcytosine in genomic DNA by sheathless interfaced capillary electrophoresis-mass spectrometry. Chem Commun 52(13):2698–2700
28. Yuan F (2020) Methodological development and application of cytosine modifications based on mass spectrometry and single-base resolution analysis. Dissertation for Ph.D in Peking University, advisors: Zhang X-X and Zhou Y-L

Chapter 4

Simple Quantification of Epigenetic DNA Modifications and DNA Damage on Multi-Well Slides

Yael Michaeli, Sapir Margalit, Sigal Avraham, Hila Erez, Noa Gilat, Zuzana Tulpová, and Yuval Ebenstein

Abstract

Chemical modifications of DNA bases have a major effect on the execution of the DNA code. The global amount of DNA modifications provides valuable information regarding various biological processes as well as for disease development. Therefore, development of simple and reliable methods to quantify these markers is of great importance. Here we describe in detail protocols for global quantification of DNA modifications. Specifically, we describe quantification of two types of epigenetic modifications, unmethylated CpGs and 5-hydroxymethylcytosine (5-hmC), and two types of DNA damage lesions, oxidation and UV-induced damage. All methods are based on utilizing enzymatic recognition for covalent binding of a fluorescent dye to the DNA modification. Up to 90 labeled DNA samples are then loaded on a custom multi-well slide, which is imaged by a conventional slide scanner. The global amount of the measured modification can be calculated by the obtained fluorescence intensity.

Keywords 5-Methylcytosine, 5-Hydroxymethylcytosine, Single-strand DNA damage, Oxidation DNA damage, UV DNA damage, Single-strand breaks, DNA modifications, Multi-well slide, Fluorescent labeling

1 Introduction

DNA modifications are native chemical changes to the molecular structure of DNA bases. Some of these modifications, such as epigenetic marks, have distinct biological function while others are the result of chemical interaction caused by environmental or endogenous exposures. The most abundant and studied epigenetic modification is methylation of the fifth position of cytosine in the context of CpG dinucleotides, known as the epigenetic mark 5-methylcytosine (5-mC). DNA methylation has a critical role in regulation of gene expression through various mechanisms, affecting many biological processes, including development, differentiation, and aging. Importantly, changes in DNA methylation contribute to the development of many types of diseases, such as

Bi-Feng Yuan (ed.), *DNA Modification Detection Methods*, Springer Protocols Handbooks,
https://doi.org/10.1007/978-1-0716-1229-3_4,

neurological disorders, immune system-related disorders, and cancer [1–3]. In 2009, the oxidation product of 5-mC, 5-hydroxymethylcytosine (5-hmC), was discovered in mammals [4]. It is the first step in the process of 5-mC oxidation and its replacement with cytosine. During the past decade, 5-hmC was the subject of many studies; yet, its role has not been completely resolved. In addition to being the intermediate form of active demethylation, 5-hmC is suggested to be important for development, differentiation, and gene expression regulation [5–7]. Its global levels are generally lower than the levels of 5-mC, however, they vary between different tissue types, ranging from as high as ~0.2% of total nucleotides in the brain, to ~0.004% in blood cells [8–10]. A significant decrease in the levels of 5-hmC was observed in many types of cancer, implying for its potential role as a diagnostic biomarker [9, 11–15].

DNA modifications also include unwanted chemical changes, such as DNA damage lesions. DNA damage can be formed endogenously due to the generation of reactive oxygen species (ROS) during the natural metabolic processes or by exposure to exogenous agents such as UV light, different types of ionization radiation, and chemicals. Although less lethal than double-strand breaks, the most common DNA damage types are single-stranded, and include abasic sites, oxidation damage, and single-strand breaks [16]. Damage levels and repair capacity were linked to disease development, progression, and treatment outcome. Additionally, it has been shown that the ability of cancer cells to repair drug-induced DNA damage can in some cases impact the efficiency of the treatment and therefore should influence the choice of therapeutic approach [17, 18].

Simple and accurate methods to measure DNA modifications are valuable for understating basic biological mechanisms as well as for disease diagnostics and treatment. Most of the current methods for measuring DNA modifications are based on antibody recognition and the efficiency and specificity of the assay are dependent on the choice of antibody. In addition, these methods are based on non-covalent binding, and thus suffer from relatively low sensitivity and sometimes poor reproducibility [9]. Measuring DNA modification by mass spectrometry analysis shows higher accuracy and sensitivity [12, 19], but requires expensive equipment, expertise, and large quantities of DNA for ultimate results.

Here we describe a method for global quantification of DNA modifications based on native chemo-enzymatic recognition. We utilize enzymes that specifically recognize different DNA modifications for covalent attachment of fluorescent dyes. DNA is then loaded on a custom multi-sample array slide, and the global amount of each modification is calculated based on the measured fluorescence intensity.

This chapter is divided into three parts. The first section describes the preparation of multi-well slides for quantification of various DNA modifications. The following sections describe the procedure for quantification of epigenetic marks and DNA damage lesions. The last section describes the generation of control samples, slide loading, imaging, and analysis. Specifically, we will describe:

(a) Quantification of unmodified CpGs (Fig. 1a).

For labeling of unmodified CpGs, we use M.TaqI methyltransferase fed with a synthetic cofactor that contains a fluorescent dye. M.TaqI labels the adenine in TCGA sequence motifs. The motif contains a nested CpG site, and labeling is blocked when this CpG is methylated. Only unmethylated CpGs result in fluorescent signal, allowing the quantification of unmodified CpGs (Fig. 1a). The enzyme can also be fed with a synthetic cofactor that contains an azide moiety, which can later be coupled via copper-free click chemistry with a strained alkyne such as dibenzocyclooctyl (DBCO) that is attached to a contrast agent of choice.

As in the gold standard bisulfite conversion method, this labeling scheme cannot distinguish between 5-mC and 5-hmC. However, since the level of methylation is significantly higher than the levels of 5-hmC or other DNA modifications, in most cases an unlabeled cytosine results from 5-mC.

Of note, M.TaqI recognizes only 5.5% of CpGs in the human genome and shows good correlation with bisulfite sequencing over 10.3% of the genome. However, it represents ~50% of gene promoters, gene bodies, and CpG islands and therefore it provides valuable information regarding the status of the functional methylome and it reliably captures differences in global genome-wide methylation levels [20].

(b) Quantification of 5-hmC (Fig. 1b).

Labeling of 5-hmC is based on the enzyme β-glucosyltransferase from the T4 phage (T4-BGT). Its natural activity is to add a glucose moiety to 5-hmC, and by that the phage protects its DNA from bacterial nucleases. We utilize this enzyme to specifically recognize 5-hmC but instead of its native cofactor, UDP-glucose, the enzyme is fed with a modified cofactor UDP-Glucose-azide (UDP-N_3-Glu). Next, the azide is labeled by a fluorescently labeled DBCO molecule via copper-free click chemistry [21–24].

(c) Quantification of oxidation DNA damage (Fig. 1c).

Oxidation DNA damage is labeled using human oxoguanine glycosylase 1 (hOGG1), which recognizes and excises oxidation damage lesions. Once the damaged base is removed, DNA polymerase that is fed with fluorescent nucleotides is used, followed by addition of a ligase that seals the strand break. As a result of this process, each damage site is replaced by a fluorescent spot.

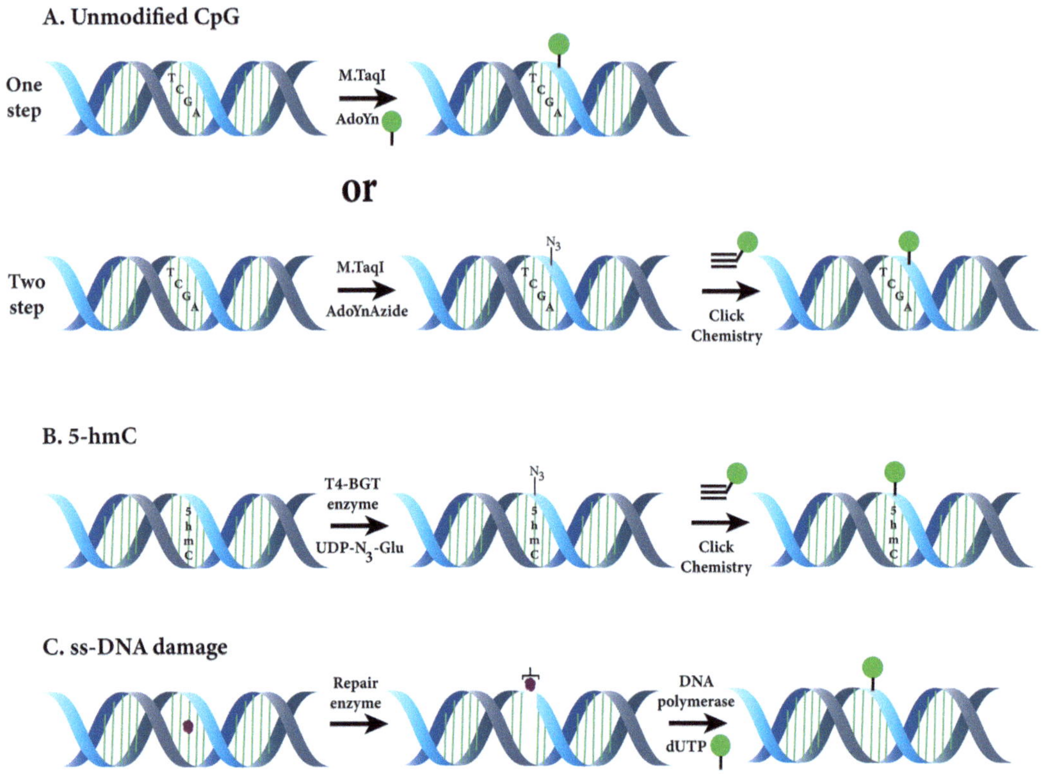

Fig. 1 Principle of DNA modifications labeling. (**a**) Labeling of unmodified CpGs. (**b**) Labeling of 5-hmC. (**c**) Labeling of DNA damage. Purple dot represents DNA damage lesion. Green dot represents a fluorophore

(d) Quantification of UV-induced DNA damage (Fig. 1c).

Similarly to quantification of oxidation damage sites, pyrimidine dimer glycosylase (T4-PDG) is used to recognize, excise, and subsequently quantify UV-induced damage sites. Of note, the signal intensity may be tuned by modifying the time of the polymerase fill-in reaction. It is thus important to use identical incubation times for the different samples to be compared.

The specific assays presented here were developed for quantification of oxidation and UV-induced damage. However, the assay may be applied to any type of single strand damage by changing the recognition enzyme. It can also measure the overall damage state of the cell, by using an enzymatic cocktail that recognizes various types of damage [17]. Importantly, our procedure for labeling DNA damage also enables monitoring repair dynamics, by measuring damage levels at different time points following natural DNA repair [25].

Notably, all protocols presented here describe labeling of one type of DNA modification with a green fluorophore. However, the choice of fluorophore can vary according to the experimental requirements. In addition, multiplexed labeling of two types of DNA modifications, each type with a separate color, may be performed on the same DNA sample.

2 Materials

2.1 General Reagents

EvaGreen DNA binding dye, 1.25μM diluted in 90% water, and 10% DMSO (Biotium).

Ultrapure water.

10× NEBuffer 4 (New England Biolabs, NEB).

2.2 Kits for DNA Extraction and Purification

GenElute Mammalian Genomic DNA Miniprep Kit (Sigma).

E.Z.N.A™ SQ Blood DNA kit (Omega Bio-tek).

Oligo Clean and Concentrator kit (Zymo Research).

2.3 Reagents for Slides Preparation

Teflon-coated microscope slides, with customized well formation, 2 mm diameter wells, 90 wells per slide (Tekdon).

0.005% poly-L-lysine solution in water (Prepared fresh from 0.1% stock, Sigma).

PBS (Sigma), autoclaved.

PBST (PBS with 0.05% Tween 20), autoclaved.

5% w/v bovine serum albumin (Sigma) in PBS.

Nitrogen gas.

2.4 Reagents for Unmodified CpGs Labeling

TaqI methyltransferase (M.TaqI), 10 units/μL (NEB).

10× CutSmart buffer (NEB).

Synthetic cofactors: AdoYnAzide/AdoYnTAMRA (homemade [15, 20] can be supplied by the authors upon request, *see* **Note 1**).

Proteinase K, 20 mg/mL (Sigma).

Dibenzocyclooctyl PEG4-5/6-TAMRA (DBCO-TAMRA), 10 mM, dissolved in DMSO (Jena Bioscience, *see* **Note 2**).

2.5 Reagents for 5-hmC Labeling

T4-β glucosyltransferase (T4-BGT), 10 units/μL (NEB).

Uridine diphosphate-6-azideglucose (UDP-6-N3-Glu), 3 mM (homemade [22], can also be purchased from Jena Bioscience, *see* **Note 1**).

Dibenzocyclooctyl PEG4-5/6-TAMRA (DBCO-TAMRA), 10 mM, dissolved in DMSO (Jena Bioscience, *see* **Note 2**).

2.6 Reagents for Labeling Oxidation Damage and UV-Induced Damage

Bovine serum albumin solution, 1 mg/mL (NEB).

Endonuclease IV, 10 units/μL (NEB).

b-Nicotinamide adenine dinucleotide (NAD+), 50 mM (NEB).

Deoxynucleotides dATP, dGTP, dCTP (sigma), 10μM each.

ATTO-550-dUTP (Jena biosciences GMBH), 10μM.

Bst DNA polymerase, large fragment, 8 units/μL (NEB).

Taq DNA ligase, 40 units/μL (NEB).

For labeling of oxidation damage: human recombinant 8-oxoguanine DNA glycosylase (hOGG1), 0.5 mg/mL (ProSpec-Tany Technogene Ltd.).

For labeling of UV-induced damage: T4 Endonuclease V (T4-PDG), 10 units/μL (NEB).

3 Methods

3.1 Summary of the Workflow

The general workflow contains the following steps (Fig. 2):

1. DNA extraction.
2. Slide preparation.
3. Fluorescent labeling of DNA modifications.
4. Applying labeled DNA samples on an activated slide.
5. Slide imaging for DNA modifications.
6. Total DNA staining.
7. Slide imaging for total DNA.
8. Data analysis.

Note: DNA extraction can be performed with a method of choice. All the kits mentioned in Subheading 2 were checked and found to be suitable for the method. See Subheading 4 for quality requirement of the extracted DNA (**Notes 3–4**).

3.2 Preparation of Multi-Well Slides

1. In a 50 mL tube or a custom slide holder, immerse Teflon-coated microscope slides with customized wells in 0.005% poly-L-lysine solution in order to positively charge the surface (*see* **Note 5**). In case 50 mL tubes are used, only one slide should be immersed in each tube, in a volume of 50 mL. Incubate for 1 h at 37 °C with light shaking (25 rpm) and then overnight at 4 °C with no shaking. Wash slides twice with PBST solution and twice with PBS by shaking manually for a few seconds and discarding the wash buffer.
2. Blocking: In a 50 mL tube, a custom slide holder or a petri dish, immerse slides in 5% w/v bovine serum albumin solution in PBS. Incubate the immersed slides for 1 h at 37 °C with light

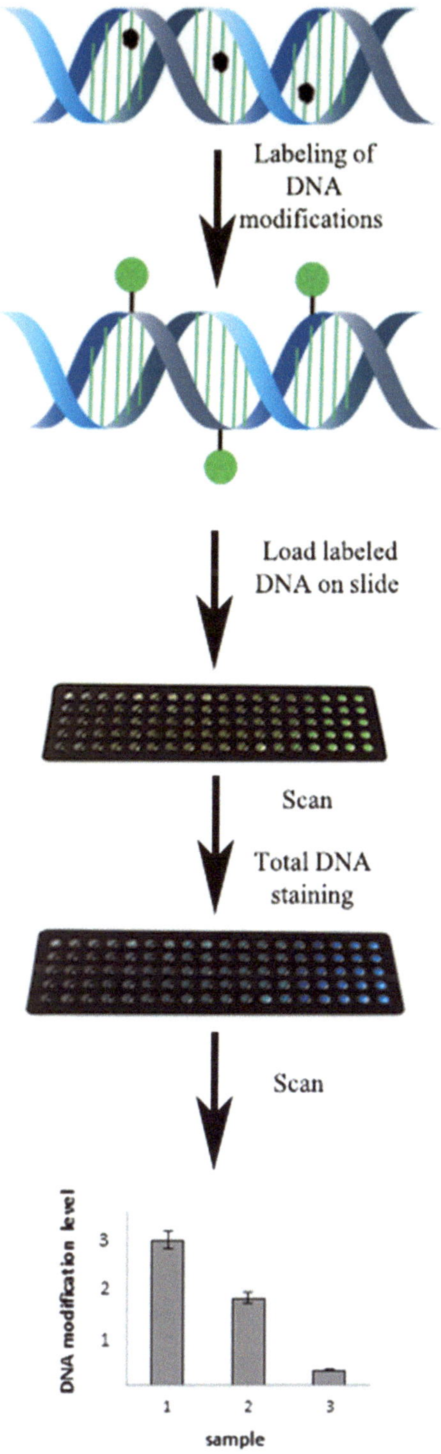

Fig. 2 Summary of the workflow

shaking (25 rpm) and then overnight at 4 °C with no shaking (*see* **Note 6**).

3. Wash slides twice with PBS and then three times with ultrapure water. Hold slide, wash briefly with a squeeze bottle filled with ultrapure water, and dry under a flow of nitrogen gas. The slides should be used immediately upon drying (*see* **Note 7**).

3.3 Labeling of Unmodified CpGs

Labeling is performed on purified DNA. In general, an experiment should contain the sample of choice, a control sample, and a calibration sample. Control samples undergo the same procedure without the addition of the M.TaqI enzyme, and should be prepared for subtraction of background signal during the analysis. For quantitative measurements, a calibration sample with known amount of unmodified CpGs in the context of TCGA should be prepared. All samples should be labeled side by side to account for variations arising from local experimental parameters. See Subheading 3.7 for preparation of calibration and control samples.

1. To a 1.5 mL tube add: 500 ng DNA, 2.5μL 10× CutSmart buffer, a modified cofactor AdoYnAzide (for a two-step reaction) or AdoYnTAMRA (for a one-step reaction) in a final concentration of 40μM, 2μL (20 units) of M.TaqI and ultrapure water to a final volume of 25μL. Mix by flicking the tube, spin down briefly, and incubate 1 h at 60 °C.

   Note: the synthetic cofactor should be thawed on ice. M. TaqI should be added last.

2. Add 2μL of proteinase K and incubate 2 h at 45 °C. This step is performed to remove M.TaqI enzymes that are bound to the DNA.

3. In case an AdoYnAzide cofactor was used in **step 1**: add 0.625μL DBCO-TAMRA (final concentration of 250μM). Mix by flicking the tube, spin down briefly, and incubate overnight at 37 °C (*see* **Note 1**).

   In case a fluorescent cofactor (AdoYnTAMRA) was used, continue directly to the purification step (**step 4**).

4. Purify DNA with "Oligo Clean and Concentrator kit" according to manufacturer's recommendations with three washing steps (*see* **Note 8**). Keep on 4 °C until analyzed.

3.4 Labeling of 5-hmC

Labeling is performed on purified DNA. In general, an experiment should contain the sample of choice, a control sample, and a calibration sample. Control samples undergo the same procedure without the addition of T4-BGT enzyme, and should be prepared for subtraction of background signal during the analysis. For quantitative measurements, a calibration sample with known amount of 5-hmC should be prepared. All samples should be labeled side by side to account for variations arising from local experimental parameters. See Subheading 3.7 for preparation of calibration and control samples.

Note: Since 5-hmC is generally less abundant than unmodified CpGs, a larger starting amount of DNA is used for labeling to allow for sufficient signal. This amount may be tuned according to the levels of 5-hmC in the analyzed sample.

1. To a 1.5 mL tube add: 1μg DNA, 3μL 10× NEBuffer 4, 0.45μL UDP-6-N_3-Glu to a final concentration of 45μM, 2μL (20 units) of T4-BGT, and ultrapure water to a final volume of 30μL. Mix by flicking the tube, spin down briefly, and incubate overnight at 37 °C.

   Note: UDP-6-N_3-Glu should be thawed on ice (*see* **Note 1**). T4-BGT should be added last.
2. Add 0.45μL DBCO-TAMRA (final concentration of 150μM). Mix and incubate overnight at 37 °C (*see* **Note 2**).
3. Purify DNA with "Oligo Clean and Concentrator" kit according to manufacturer's recommendations with three washing steps (*see* **Note 8**). Keep on 4 °C until analyzed.

3.5 Labeling of Oxidation Damage

Labeling of oxidation damage is performed in three consecutive enzymatic reactions. A control sample with no repair enzyme should be prepared for subtraction of background signal during the analysis. An additional sample with no induced damage should be prepared and labeled side by side with the test sample, for measurement of the basal damage level.

1. To a 1.5 mL tube add: 500 ng of DNA sample, 1.5μL of 10× NEBuffer 4, 1.5μL bovine serum albumin solution (1.5μg), 0.3μL of hOGG1 (0.15μg), and ultrapure water to a final volume of 15μL. Mix by flicking the tube, spin down briefly, and incubate for 30 min at 37 °C.
2. Add 0.5μL of Endonuclease IV (5 units) enzyme to the reaction, mix and incubate for additional 30 min at 37 °C.
3. Add the following to the reaction tube: 1.5μL of 10× NEBuffer 4, 1.5μL bovine serum albumin solution (1.5μg), 0.2μL NAD+ (to a final concentration of 0.33 mM), 0.3μL of each of the following nucleotides: dATP, dGTP, dCTP, and fluorescent ATTO-550-UTP (to a final concentration of 100 nM each), 0.4μL of Bst DNA polymerase, large fragment (3.2 units), 0.2μL Taq DNA ligase (8 units), and ultrapure water to a final volume of 30μL. Mix and incubate for 30 min at 65 °C.
4. Purify the labeled DNA samples from excess fluorescent dyes using "Oligo Clean and Concentrator" columns, according to manufacturer's recommendations, with one additional washing step (*see* **Note 8**).

3.6 Labeling of UV-Induced DNA Damage

Labeling reaction of photoproducts is identical to labeling of oxidation damage, apart from using 0.5μL of T4-PDG (5 units) in **step 1**, instead of using hOGG1 (*see* **Note 9**).

3.7 Preparation of Negative Control and Calibration Samples

3.7.1 Calibration Samples

In case an absolute level of DNA damage or epigenetic marks is desired, calibration samples with known quantity of the tested modification should be used. Two to four such samples may be used to construct a calibration curve. The calibration samples should contain the tested modification at levels equivalent to those of the analyzed DNA for reliable interpolation from the calibration curve. The sample can be synthetically prepared with known amounts of the modification, amplified with modified nucleotides, or, alternatively, the level of modification can be determined by LC–MS/MS [15, 19]. The calibration samples and the tested samples should be labeled and loaded on the slide side by side in each experiment, to account for variations in the labeling and loading procedures between experiments. The stock of these standards can be prepared once and can be used for multiple slides.

3.7.2 Control Samples

Control samples are generated by an identical process as the test samples but without the labeling enzyme. The fluorescence signal measured in these samples originates from the remaining free fluorophores. This background signal will be subtracted during analysis to reliably represent the signal derived solely from labeled DNA modifications.

For labeling of DNA damage, an additional control sample is composed of DNA with no induced damage. This sample represents the basal damage level and the relative damage can be calculated according to this level.

3.8 Loading of Labeled DNA on Slides

1. Use a slide prepared according to Subheading 3.2. Load 1μL purified, labeled DNA sample in each well. For optimal attachment, 10–30 ng of DNA should be loaded per well. Four to five replicates of each sample should be loaded for optimal results (Fig. 3).
2. Load four to five wells with 1μL control sample as described above.
3. In case used, load four to five wells with 1μL calibration sample as described above.
4. Load four to five wells with 1μm ultrapure water to account for background signal of DNA-binding dye (see Subheading 3.12).
5. Incubate slides for 14 min at 42 °C and then for 24 min at 30 °C in humid conditions by applying water drops around the slides, to avoid rapid drying of the wells.

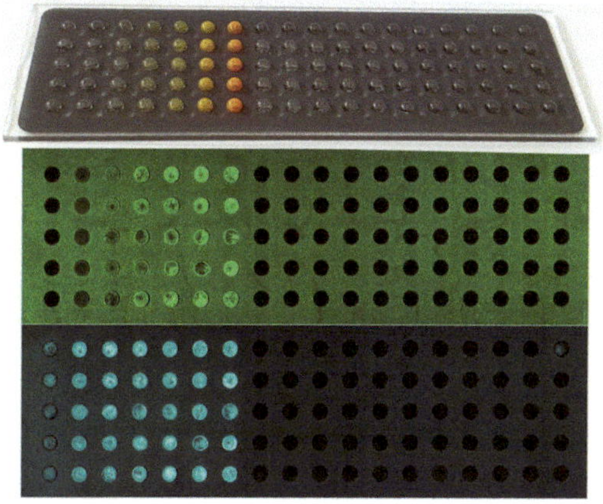

Fig. 3 A typical multi-well slide after sample loading (top) and after imaging of DNA modifications (green, middle) and total DNA (blue, bottom)

6. Wash slides once with ultrapure water in a petri dish by shaking manually for a few seconds. Hold slide gently, wash briefly with a squeeze bottle filled with ultrapure water, and dry under a flow of nitrogen gas.

   Keep slides in the dark until imaging to avoid photobleaching.

3.9 Imaging of Fluorescent Labels

Image slides using InnoScan1100 slide scanner (Innopsys) or equivalent (Fig. 3, middle). A 532 nm green laser should be used to image the ATTO-550 or TAMRA fluorophores. Make sure the fluorescent signal is not saturated. In case it is, reduce the laser power or exposure time and image again.

Note: The ATTO-550/TAMRA labels should be imaged before total DNA staining by EvaGreen to avoid their co-excitation by the green laser.

3.10 Total DNA Quantification

1. To stain total DNA with EvaGreen DNA binding dye, add 1μL of 1.25μM dye to each well containing DNA. In addition, stain five wells containing only water with no DNA. These wells will be used to calculate the background signal of the EvaGreen dye in the absence of DNA.
2. Incubate slides for 30 min at room temperature in the dark.
3. Wash slides once with ultrapure water in a petri dish by shaking manually for a few seconds. Hold slide gently, wash briefly with a squeeze bottle filled with ultrapure water, and dry under a flow of nitrogen gas.

3.11 Imaging of Total DNA

Image slides using InnoScan1100 slide scanner (Innopsys) or equivalent (Fig. 3, bottom). A 488 nm blue laser should be used to image the EvaGreen stain. Make sure the fluorescent signal is not saturated. In case it is, reduce the laser power or exposure time and image again.

3.12 Analysis

Analyze images using ImageJ. Extract the mean fluorescence intensity of the EvaGreen (total DNA) and the ATTO-550/TAMRA (epigenetic/DNA damage labels) from each well. Determine the mean background signal of the ATTO-550/TAMRA from the control wells (wells without the labeling enzymes). Subtract the mean background signal from the ATTO-550/TAMRA fluorescence signal in each sample well. Determine the background signal of the EvaGreen (total DNA), by calculating the mean fluorescence signal of all wells containing EvaGreen and no DNA. Subtract the background EvaGreen signal from the EvaGreen fluorescence signal in each sample well. Following background subtractions in both channels, divide the calculated ATTO-550/TAMRA signal in each well by the calculated signal in the EvaGreen channel of the same well. This step is will normalize the signal to the amount of DNA in the well. Finally, for each sample, calculate the average and standard deviation over four to five replicates. This analysis determines the relative quantity of epigenetic marks/DNA damage lesions. An equation for calculation the relative amount of DNA modification is presented below:

$$\text{Relative DNA modification value (exp.)} = \frac{\sum_{1}^{n} \frac{\text{TAMRA or ATTO550 fluorescence signal (well with labeled sample} - \text{mean of control)}}{\text{EvaGreen fluorescence signal (well of labeled sample} - \text{EvaGreen background noise)}}}{n}$$

In case a calibration sample with known percentage of DNA modification was used, the absolute level of the modification can be calculated in each well according to the following equation (presented for calculation of 5-hmC in this case) [15]:

$$\% \frac{\text{5hmC}}{\text{total dNTPs}} = \text{sample's relative 5hmC value (exp.)} * \frac{\text{calibration sample's known 5hmC level (LC} - \text{MS/MS)}}{\text{calibration sample's relative 5hmC value (exp.)}}$$

4 Notes

1. Thaw all cofactors slowly on ice. Divide to small aliquots and avoid repeated freezing and thawing cycles.
2. Store DBCO fluorophores at −80 °C. Divide to small aliquots and avoid repeated freezing and thawing cycles.
3. Following DNA extraction, the purity of the DNA should be checked in a UV–VIS spectrometer according to 260/280 and 260/230 ratios. Bad quality DNA will result in poor adsorption to the slide.
4. The length of the DNA influences the efficiency of its adsorption to the slide. Therefore, make sure that the length of the

DNA is similar in samples that should be compared. In addition, do not use DNA fragments shorter than 200 bp, as it shows poor adsorption to the slide.

5. The shelf life of the Teflon-coated microscope slides is around 18 months. Store unused slides in dry conditions in a desiccator at all times. Keep away from dirt and dust. Dirty or expired slides result in a false fluorescent signal in the blue channel. Dry under a flow of nitrogen gas prior to initial use.
6. Prepared slides with poly-L-lysine can be kept in the blocking solution up to 2 days before using the slide.
7. Make sure to load the slides immediately after blocking and drying.
8. Several purification methods were examined for the post-labeling step. We have found that the most consistent and reliable results are obtained with Oligo Clean and Concentrator kit (Zymo Research), with two to three washing steps. To improve the yield, a second elution step can be performed. The recommended volumes for purification of 1μg DNA are 12–15μL for the first elution and 8–10μL for the second elution. Do not load more than 2μg per column.
9. The UV-induced damage reaction labels some oxidation damage in addition to photoproducts. For labeling of photoproducts only, the oxidation damage lesions should be labeled or blocked first. For more details, see Torchinsky et al. [25].

Acknowledgments

The authors acknowledge financial support from the BeyondSeq Consortium (EC program 634890), the European Research Council Proof of Concept grant by the EU-Horizon2020 program (grant no. 767931), the European Research Council Consolidator grant (grant No. 817811), the NIH R21 grant (R21ES028015-011), and the Joint Israeli German R&D Nanotechnology (grant no. 61976).

References

1. Liyanage VRB, Jarmasz JS, Murugeshan N et al (2014) DNA modifications: function and applications in normal and disease states. Biology 3:670–723
2. Jones PA (2012) Functions of DNA methylation: islands, start sites, gene bodies and beyond. Nat Rev Genet 13:484–492
3. Greenberg MVC, Bourc'his D (2019) The diverse roles of DNA methylation in mammalian development and disease. Nat Rev Mol Cell Biol 20:590–607
4. Tahiliani M, Koh KP, Shen Y et al (2009) Conversion of 5-methylcytosine to 5-hydroxymethylcytosine in mammalian DNA by MLL partner TET1. Science 324:930–935
5. Shi DQ, Ali I, Tang J et al (2017) New insights into 5hmC DNA modification: generation, distribution and function. Front Genet 8:100

6. Ficz G, Branco MR, Seisenberger S et al (2011) Dynamic regulation of 5-hydroxymethylcytosine in mouse ES cells and during differentiation. Nature 473:398–402
7. Stroud H, Feng S, Morey Kinney S et al (2011) 5-Hydroxymethylcytosine is associated with enhancers and gene bodies in human embryonic stem cells. Genome Biol 12:R54
8. Shahal T, Green O, Hananel U et al (2016) Simple and cost-effective fluorescent labeling of 5-hydroxymethylcytosine. Methods Appl Fluoresc 4:044003
9. Gilat N, Tabachnik T, Shwartz A et al (2017) Single-molecule quantification of 5-hydroxymethylcytosine for diagnosis of blood and colon cancers. Clin Epigenetics 9:70
10. Li W, Liu M (2011) Distribution of 5-hydroxymethylcytosine in different human tissues. J Nucleic Acids 2011:870726
11. Chen Z, Shi X, Guo L et al (2015) Decreased 5-hydroxymethylcytosine levels correlate with cancer progression and poor survival: a systematic review and meta-analysis. Oncotarget 8:1944–1952
12. Kroeze LI, Aslanyan MG, Van Rooij A et al (2014) Characterization of acute myeloid leukemia based on levels of global hydroxymethylation. Blood 124:1110–1118
13. Liu C, Liu L, Chen X et al (2013) Decrease of 5-hydroxymethylcytosine is associated with progression of hepatocellular carcinoma through downregulation of TET1. PLoS One 8:e62828
14. Yang Q, Wu K, Ji M et al (2013) Decreased 5-hydroxymethylcytosine (5-hmC) is an independent poor prognostic factor in gastric cancer patients. J Biomed Nanotechnol 9:1607–1616
15. Margalit S, Avraham S, Shahal T et al (2020) 5-Hydroxymethylcytosine as a clinical biomarker: fluorescence-based assay for high-throughput epigenetic quantification in human tissues. Int J Cancer 146:115–122
16. Abbotts R, Wilson DM (2017) Coordination of DNA single strand break repair. Free Radic Biol Med 107:228–244
17. Zirkin S, Fishman S, Sharim H et al (2014) Lighting up individual DNA damage sites by in vitro repair synthesis. J Am Chem Soc 136:7771–7776
18. Gavande NS, Vandervere-Carozza PS, Hinshaw HD et al (2016) DNA repair targeted therapy: The past or future of cancer treatment? Pharmacol Ther 160:65–83
19. Shahal T, Koren O, Shefer G et al (2018) Hypersensitive quantification of global 5-hydroxymethylcytosine by chemoenzymatic tagging. Anal Chim Acta 1038:87–96
20. Sharim H, Grunwald A, Gabrieli T et al (2019) Long-read single-molecule maps of the functional methylome. Genome Res:1–38
21. Michaeli Y, Shahal T, Torchinsky D et al (2013) Optical detection of epigenetic marks: sensitive quantification and direct imaging of individual hydroxymethylcytosine bases. Chem Commun (Camb) 49:8599–8601
22. Nifker G, Levy-Sakin M, Berkov-Zrihen Y et al (2015) One-pot chemoenzymatic cascade for labeling of the epigenetic marker 5-Hydroxymethylcytosine. Chembiochem 16:1857–1860
23. Jain N, Shahal T, Gabrieli T et al (2019) Global modulation in DNA epigenetics during pro-inflammatory macrophage activation. Epigenetics 14:1183–1193
24. Gabrieli T, Sharim H, Nifker G et al (2018) Epigenetic optical mapping of 5-hydroxymethylcytosine in nanochannel arrays. ACS Nano 12:7148
25. Torchinsky D, Michaeli Y, Gassman NR et al (2019) Simultaneous detection of multiple DNA damage types by multi-colour fluorescent labelling. Chem Commun 55:11414–11417

Chapter 5

A Label-Free Electrochemical Biosensor for Sensitive Detection of 5-Hydroxymethylcytosine

Lin Cui, Juan Hu, Meng Wang, Chen-Chen Li, and Chun-Yang Zhang

Abstract

The 5-hydroxymethylcytosine (5-hmC) is regarded as sixth base in the genome and closely linked to a series of physiological and pathological processes. Nevertheless, the sensitive detection of 5-hmC and efficient discrimination of 5-hmC from 5-methylcytosine (5-mC) remain challenging. To solve these issues, we constructed a novel electrochemical biosensor for 5-hmC assay on the basis of terminal deoxynucleotidyl transferase (TDT)-mediated signal amplification and Ru (III) redox cycling. The proposed biosensor is simple and low cost without the involvement of any immobilization and labeling steps, and it achieves high sensitivity with a detection limit down to 9.06 fM. Moreover, it possesses high specificity with the capability of discriminating target 5-hmC from 5-mC, and it has been successfully applied for the detection of endogenous 5-hmC in human cell lines.

Keywords 5-Hydroxymethylcytosine, Electrochemical magnetobiosensor, Ru (III) redox cycling, Terminal deoxynucleotidyl transferase, SPCE, Label-free, Immobilization-free

1 Introduction

The 5-hydroxymethylcytosine (5-hmC) is a ten–eleven translocation (TET) enzymes-catalyzed oxidant product of 5-methylcytosine (5-mC) and has been regarded as the sixth base besides A, T, C, G, and 5-mC [1]. Like 5-mC, the 5-hmC also acts as a potential epigenetic marker and plays crucial roles in diverse physiological processes such as DNA demethylation [2], transcriptional regulation [3], epigenetic reprogramming [4], and embryonic development [5–7]. Moreover, the global decrease of 5-hmC content is closely linked to human cancers [8, 9], and the 5-hmC signatures in circulating DNA may function as predictive biomarkers for colorectal and gastric cancers [10]. Therefore, the efficient detection of 5-hmC is of great importance to 5-hmC-related epigenetic researches and clinical disease diagnosis.

The bisulfite sequencing is the gold standard for 5-mC detection [11], but it is unable to discriminate 5-hmC from 5-mC. In

Bi-Feng Yuan (ed.), *DNA Modification Detection Methods*, Springer Protocols Handbooks,
https://doi.org/10.1007/978-1-0716-1229-3_5,

contrast, the methods based on mass spectrometry [12], chromatography [13–15], radioactive labeling [16], single-molecule sequencing [17], and immunoassay [18–20] possess high selectively that enables the discrimination of 5-hmC from 5-mC. Nevertheless, the wide applications of these methods are hampered by expensive and hazardous labels, sophisticated instruments, laborious procedures, and unsatisfactory sensitivity. Consequently, there is an urgent need for the development of simple, selective, and sensitive 5-hmC assay.

The electrochemical biosensors exhibit significant advantages of portability, low cost, and high sensitivity [21, 22] which has been successfully applied for 5-hmC assay in combination with enzymatic modification [23, 24]. Nevertheless, these biosensors suffer from expensive enzymes, strict assay conditions, and laborious synthesis and modification protocols. To eliminate the usage of the enzyme, the electrogenerated chemiluminescence (ECL) biosensor on the basis of $KRuO_4$ oxidation and ECL labeling [25] and nanopore biosensor based on bisulfite-mediated biotinylation [26] are well established. However, applications of these methods for 5-hmC detection in genomic DNA have not been reported.

Herein, a novel electrochemical biosensor is constructed for sensitive detection of 5-hmC in genomic DNA on the basis of terminal deoxynucleotidyl transferase (TDT)-mediated signal amplification and Ru(III) redox cycling. The proposed biosensor achieves high sensitivity with a detection limit of as low as 9.06 fM without any immobilization steps [27, 28], and it can further be used for the detection of 5-hmC in human cancer cells and the discrimination of 5-hmC from cytosine and 5-mC.

The principle of the proposed electrochemical biosensor is shown in Fig. 1. The 5-hmC ssDNA can hybridize with complementary DNA to form a dsDNA, and the 5-hmC is modified with biotin by using a biotinyl–cysteine derivative (Fig. 1a). The biotinylated dsDNAs are then captured by streptavidin-conjugated magnetic bead (MB) via specific biotin–streptavidin interaction (Fig. 1b), and the 3′-hydroxy termini of complementary DNA is extended by TDT to form longer DNA strands. Because the DNAs are negatively charged, the $Ru(NH_3)_6^{3+}$ can be attached to the newly elongated DNA strands on MBs through electrostatic interaction and further brought close to the screen-printed carbon electrode (SPCE) surface under magnetic field. With $Fe(CN)_6^{3-}$ as secondary electron acceptor, amplified electrocatalytic current can be generated according to following electron transfer kinetic-based mechanism:

$$Ru(NH_3)_6^{3+} + e^- \rightarrow Ru(NH_3)_6^{2+} \quad (1)$$

$$Ru(NH_3)_6^{2+} + Fe(CN)_6^{3-} \rightarrow Ru(NH_3)_6^{3+} + Fe(CN)_6^{4-} \quad (2)$$

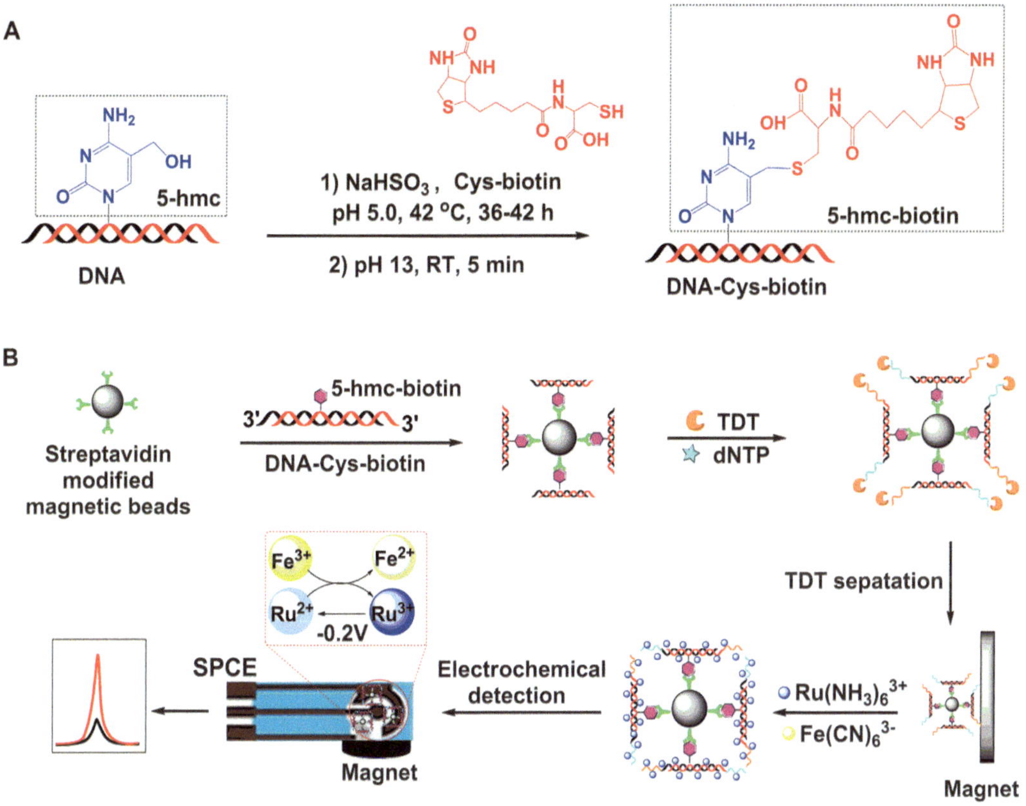

Fig. 1 Principle of the electrochemical 5-hmC biosensor. Two main steps are involved: (**a**) single-step bisulfite-mediated biotinylation of 5-hmC and (**b**) dual signal amplification based on TdT polymerization and Ru(III) redox recycling. (Reproduced from ref. [29] with permission from American Chemical Society)

Notably, because the negatively charged $Fe(CN)_6^{3-}$ is unable to access the SPCE surface masked by the partially anionic DNA, the false-positive signal can be eliminated when target is absent. Owing to the TDT-mediated polymerization and Ru(III) redox recycling-assisted dual signal amplification, the proposed biosensor enables selective and ultrasensitive detection of 5-hmC.

2 Materials

2.1 Solutions

1. Electrochemical signals were measured in 1 mM PBS (0.1 M NaCl, pH 7.4) containing 2 mM $Fe(CN)_6^{3-}$ and 27 μM Ru $(NH_3)_6^{3+}$ (*see* **Note 1**).
2. Cys-biotin (10 μM, pH was adjusted to 5.0 with 10 M NaOH).

3. Electrochemical impedance spectroscopic (EIS) measurements were performed in 5 mM $[Fe(CN)_6]^{3-/4-}$ redox couple solution (1:1 molar ratio) containing 0.1 M KCl.
4. $Na_2S_2O_5$ (1.0 M, pH 5.0).
5. Washing buffer (1 mM EDTA, 20 mM Tris–HCl, pH 7.4, 0.5 M NaCl).
6. 1× TdT buffer (50 mM KAc, 20 mM Tris-Ac, 10 mM Mg $(Ac)_2$, pH 7.9).
7. 0.25 mM $CoCl_2$.
8. 1 mM dNTPs.
9. 10% fetal bovine serum.

2.2 Equipment

1. All electrochemical measurements were carried out at room temperature on a CHI 660e electrochemical workstation (CH Instruments Inc., USA) with a portable homemade screen-printed carbon electrode (SPCE), which consisted of an Ag/AgCl reference electrode, a carbon auxiliary electrode, and a carbon working electrode (2 mm diameter).
2. A Malvern Zeta Sizer Nano (Malvern Instruments, USA) was used to measure Zeta potential.

2.3 Enzymes, Kit, and Other Materials

1. Terminal deoxynucleotidyl transferase (TdT).
2. Micro-Bio-Spin P6 column was purchased from New England Biolabs (NEB, UK).
3. Streptavidin-coated magnetic bead (MB) (1 μm, 10 mg mL^{-1}, Dynabeads MyOne™ Streptavidin, C1) (*see* **Note 2**).
4. Screen-printed carbon electrode (SPCE) (*see* **Note 3**).

3 Methods

3.1 Preparation of Biotinylated 5-hmC DNA

1. The biotinylated dsDNAs (10 μL) were prepared by incubating 10 nM complementary DNA with different concentration of 5-hmC DNA. Cys-biotin (120 μL, 10 μM, pH was adjusted to 5.0 with 10 M NaOH).
2. 10 μL of dsDNA (i.e., 5-hmC DNA-complementary DNA) was mixed in 200 μL of Eppendorf tube.
3. Then $Na_2S_2O_5$ (5 μL, 1.0 M, pH 5.0) was added to the solution, followed by incubation for 48 h at 42 °C.
4. After cooling to room temperature, the solution was adjusted to pH 13 by using 10 M NaOH.
5. After 5 min, the excess Cys-biotin and salts were removed by passing the sample three times through a Micro-Bio-Spin P6 column (Bio Rad) (Fig. 2).

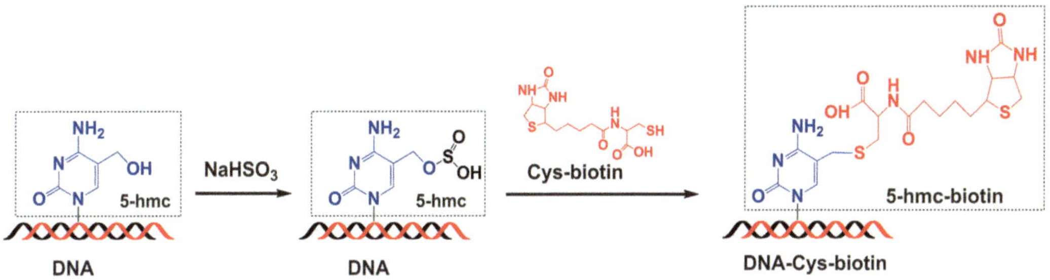

Fig. 2 Principle of the modification of 5-hmC by biotinyl-cysteine derivative. (Reproduced from ref. [29] with permission from American Chemical Society)

3.2 Preparation of MBs Linking 5-hmC DNA

1. The 1.6 μL of the streptavidin-coated MBs suspension was added to the above 200 μL of Eppendorf tube, followed by washing twice with 50 μL of washing buffer (1 mM EDTA, 20 mM Tris–HCl, pH 7.4, 0.5 M NaCl).
2. The supernatants were removed by magnet to separate the MBs.
3. Enzyme amplification was carried out in 50 μL of reaction system containing 1× TdT buffer (50 mM KAc, 20 mM Tris-Ac, 10 mM $Mg(Ac)_2$, pH 7.9), 1 mM dNTPs, 0.25 mM $CoCl_2$, 6 U of TdT, and the above separated MBs.
4. The supernatants were removed by magnet to separate the MB, followed by washing with the washing buffer to form the sensor array.

3.3 Electrochemical Measurements

1. The solution containing 2 mM $Fe(CN)_6^{3-}$, 0.1 M NaCl, 1 mM PBS (pH 7.4), and 27 μM $Ru(NH_3)_6^{3+}$ was added into the Eppendorf tube containing the separated biotinylated 5-hmC DNA/MB.
2. Then 100 μL of above solution was pipetted onto the screen-printed working electrode (SPCE) surface.
3. The super magnet underneath the working electrode surface enables the immobilization of biotinylated 5-hmC DNA/MB onto the SPCE surface.
4. Differential pulse voltammetry (DPV) in the range from 0 to −500 mV (vs. Ag/AgCl) with a pulse width of 50 ms and a pulse amplitude of 50 mV was carried out to record the electrochemical responses.

3.4 Isolation of DNA from Cell Lines

1. The cells were cultured in Dulbecco's modified Eagle's medium (DMEM) supplemented with 1% penicillin–streptomycin (Gibco, USA) and 10% fetal bovine serum (FBS, Gibco, USA) in a humidified atmosphere containing 5% CO_2 at 37 °C.
2. The cells in the exponential phase of growth were collected with trypsinization, followed by washing twice with ice-cold

PBS (pH 7.4, Gibco, USA) and centrifugation for 5 min at 200 × *g* (The DNA samples from cells were prepared using the cultured cell DNA extraction kit (Epigentek) according to the manufacturer's protocol).

3. The obtained cells were suspended in 100 μL of lysis buffer (10 mM Tris–HCl, pH 7.5, 10% glycerol, 5 mM mercaptoethanol, 0.1 mM PMSF, 1 mM EGTA, 1 mM $MgCl_2$, 0.5% CHAPS).
4. After centrifugation, the supernatant was stored at −80 °C prior to use.

3.5 Real Sample Analysis

The feasibility of electrochemical magnetobiosensor for real sample analysis is verified by measuring 5-hmC DNA in serum samples. We added different concentration of 5-hmC DNA into the 100-fold diluted serum and measured the corresponding recoveries.

4 Notes

1. The mixture of 2 mM $Fe(CN)_6^{3-}$ and $Ru(NH_3)_6^{3+}$ should be used in fresh.
2. The magnetic bead was washed prior to use and its size should be optimized.
3. The SPCE should be used once at a time.

Acknowledgments

This work was supported by the National Natural Science Foundation of China (Grant Nos. 21735003, 21527811, 21605096, and 21575152), and the Award for Team Leader Program of Taishan Scholars of Shandong Province, China.

References

1. Münzel M, Globisch D, Carell T (2011) 5-Hydroxymethylcytosine, the sixth base of the genome. Angew Chem Int Ed 50 (29):6460–6468
2. Nawy T (2013) Dynamics of DNA demethylation. Nat Methods 10(6):466–466
3. Wu H, D'Alessio AC, Ito S, Wang Z, Cui K, Zhao K, Sun YE, Zhang Y (2011) Genome-wide analysis of 5-hydroxymethylcytosine distribution reveals its dual function in transcriptional regulation in mouse embryonic stem cells. Genes Dev 25(7):679–684
4. Wossidlo M, Nakamura T, Lepikhov K, Marques CJ, Zakhartchenko V, Boiani M, Arand J, Nakano T, Reik W, Walter J (2011) 5-Hydroxymethylcytosine in the mammalian zygote is linked with epigenetic reprogramming. Nat Commun 2:241
5. Ficz G, Branco MR, Seisenberger S, Santos F, Krueger F, Hore TA, Marques CJ, Andrews S, Reik W (2011) Dynamic regulation of 5-hydroxymethylcytosine in mouse ES cells and during differentiation. Nature 473 (7347):398–402
6. Koh KP, Yabuuchi A, Rao S, Huang Y, Cunniff K, Nardone J, Laiho A, Tahiliani M, Sommer CA, Mostoslavsky G, Lahesmaa R, Orkin SH, Rodig SJ, Daley GQ, Rao A

(2011) Tet1 and Tet2 Regulate 5-hydroxymethylcytosine production and cell lineage specification in mouse embryonic stem cells. Cell Stem Cell 8(2):200–213

7. Branco MR, Ficz G, Reik W (2012) Uncovering the role of 5-hydroxymethylcytosine in the epigenome. Nat Rev Genet 13(1):7–13
8. Kudo Y, Tateishi K, Yamamoto K, Yamamoto S, Asaoka Y, Ijichi H, Nagae G, Yoshida H, Aburatani H, Koike K (2012) Loss of 5-hydroxymethylcytosine is accompanied with malignant cellular transformation. Cancer Sci 103(4):670–676
9. Lian CG, Xu Y, Ceol C, Wu F, Larson A, Dresser K, Xu W, Tan L, Hu Y, Zhan Q, Lee CW, Hu D, Lian BQ, Kleffel S, Yang Y, Neiswender J, Khorasani AJ, Fang R, Lezcano C, Duncan LM, Scolyer RA, Thompson JF, Kakavand H, Houvras Y, Zon LI, Mihm MC Jr, Kaiser UB, Schatton T, Woda BA, Murphy GF, Shi YG (2012) Loss of 5-hydroxymethylcytosine is an epigenetic hallmark of melanoma. Cell 150(6):1135–1146
10. Li W, Zhang X, Lu X, You L, Song Y, Luo Z, Zhang J, Nie J, Zheng W, Xu D, Wang Y, Dong Y, Yu S, Hong J, Shi J, Hao H, Luo F, Hua L, Wang P, Qian X, Yuan F, Wei L, Cui M, Zhang T, Liao Q, Dai M, Liu Z, Chen G, Meckel K, Adhikari S, Jia G, Bissonnette MB, Zhang X, Zhao Y, Zhang W, He C, Liu J (2017) 5-Hydroxymethylcytosine signatures in circulating cell-free DNA as diagnostic biomarkers for human cancers. Cell Res 27(10):1243–1257
11. Huang Y, Pastor WA, Shen Y, Tahiliani M, Liu DR, Rao A (2010) The behaviour of 5-hydroxymethylcytosine in bisulfite sequencing. PLoS One 5(1):e8888
12. Tang Y, Zheng S-J, Qi C-B, Feng Y-Q, Yuan B-F (2015) Sensitive and simultaneous determination of 5-methylcytosine and its oxidation products in genomic DNA by chemical derivatization coupled with liquid chromatography-tandem mass spectrometry analysis. Anal Chem 87(6):3445–3452
13. Tahiliani M, Koh KP, Shen Y, Pastor WA, Bandukwala H, Brudno Y, Agarwal S, Iyer LM, Liu DR, Aravind L, Rao A (2009) Conversion of 5-methylcytosine to 5-hydroxymethylcytosine in mammalian DNA by MLL partner TET1. Science 324(5929):930–935
14. Ito S, D'Alessio AC, Taranova OV, Hong K, Sowers LC, Zhang Y (2010) Role of Tet proteins in 5mC to 5hmC conversion, ES-cell self-renewal and inner cell mass specification. Nature 466(7310):1129–1133
15. Liutkeviciute Z, Kriukiene E, Grigaityte I, Masevicius V, Klimasauskas S (2011) Methyltransferase-directed derivatization of 5-hydroxymethylcytosine in DNA. Angew Chem Int Ed 50(9):2090–2093
16. Szwagierczak A, Bultmann S, Schmidt CS, Spada F, Leonhardt H (2010) Sensitive enzymatic quantification of 5-hydroxymethylcytosine in genomic DNA. Nucleic Acids Res 38(19):e181
17. Song C-X, Clark TA, Lu X-Y, Kislyuk A, Dai Q, Turner SW, He C, Korlach J (2012) Sensitive and specific single-molecule sequencing of 5-hydroxymethylcytosine. Nat Methods 9(1):75–77
18. Nestor CE, Ottaviano R, Reddington J, Sproul D, Reinhardt D, Dunican D, Katz E, Dixon JM, Harrison DJ, Meehan RR (2012) Tissue type is a major modifier of the 5-hydroxymethylcytosine content of human genes. Genome Res 22(3):467–477
19. Zhong S, Li Z, Jiang T, Li X, Wang H (2017) Immunofluorescence imaging strategy for evaluation of the accessibility of DNA 5-hydroxymethylcytosine in chromatins. Anal Chem 89(11):5702–5706
20. Pastor WA, Pape UJ, Huang Y, Henderson HR, Lister R, Ko M, McLoughlin EM, Brudno Y, Mahapatra S, Kapranov P, Tahiliani M, Daley GQ, Liu XS, Ecker JR, Milos PM, Agarwal S, Rao A (2011) Genome-wide mapping of 5-hydroxymethylcytosine in embryonic stem cells. Nature 473(7347):394–397
21. Wang X, Chen F, Zhang D, Zhao Y, Wei J, Wang L, Song S, Fan C, Zhao Y (2017) Single copy-sensitive electrochemical assay for circulating methylated DNA in clinical samples with ultrahigh specificity based on a sequential discrimination-amplification strategy. Chem Sci 8(7):4764–4770
22. Chen F, Wang X, Cao X, Zhao Y (2017) Accurate electrochemistry analysis of circulating methylated DNA from clinical plasma based on paired-end tagging and amplifications. Anal Chem 89(19):10468–10473
23. Chen S, Dou Y, Zhao Z, Li F, Su J, Fan C, Song S (2016) High-sensitivity and high-efficiency detection of DNA hydroxymethylation in genomic DNA by multiplexing electrochemical biosensing. Anal Chem 88(7):3476–3480
24. Zhou Y, Yang Z, Li X, Wang Y, Yin H, Ai S (2015) Electrochemical biosensor for detection of DNA hydroxymethylation based on glycosylation and alkaline phosphatase catalytic signal amplification. Electrochim Acta 174:647–652

25. Ma S, Sun H, Li Y, Qi H, Zheng J (2016) Discrimination between 5-hydroxymethylcytosine and 5-methylcytosine in DNA via selective Electrogenerated Chemiluminescence (ECL) Labeling. Anal Chem 88(20):9934–9940
26. Li W-W, Gong L, Bayley H (2013) Single-molecule detection of 5-hydroxymethylcytosine in DNA through chemical modification and nanopore analysis. Angew Chem Int Ed 52(16):4350–4355
27. Campuzano S, Torrente-Rodriguez RM, Lopez-Hernandez E, Conzuelo F, Granados R, Sanchez-Puelles JM, Pingarron JM (2014) Magnetobiosensors based on viral protein p19 for microRNA determination in cancer cells and tissues. Angew Chem Int Ed 53(24):6168–6171
28. Chen Y, Xianyu Y, Dong M, Zhang J, Zheng W, Qian Z, Jiang X (2018) Cascade reaction-mediated assembly of magnetic/silver nanoparticles for amplified magnetic biosensing. Anal Chem 90(11):6906–6912
29. Cui L, Hu J, Wang M, Li CC, Zhang CY (2019) Label-free and immobilization-free electrochemical magnetobiosensor for sensitive detection of 5-hydroxymethylcytosine in genomic DNA. Anal Chem 91(2):1232–1236

Chapter 6

Electrochemical Assay for Continuous Monitoring of Dynamic DNA Methylation Process

Zong Dai, Li Zhang, Si-Yang Liu, Yuzhi Xu, Danping Chen, Jun Chen, and Xiaoyong Zou

Abstract

DNA methylation is an important mode of epigenetic modification, which plays key roles in many cellular processes. Within the last decades, various methods have been proposed for quantitative analysis of methylated DNA. However, the reports relating to the monitoring of dynamic methylation process are rare. The challenges of tracking DNA methylation process mainly include the detection of minute change during the methylation process and the synchronously amplifying of target signals. Herein, we propose an electrochemical strategy for continuous monitoring of DNA methylation process over time based on long-range electron transfer. An electrochemical sensor is prepared by assembling single-strand DNA probes whose tops are labeled with 6-ferrocenylhexanethiol-modified gold nanoparticle. No obvious current response can be observed until the establishment of the long-range electron transfer pathway between 6-ferrocenylhexanethiol and electrode by hybridization of the complementary DNA. Once the DNA is methylated, a bromine group will be immediately colocated onto it in the presence of $NaIO_4$/LiBr. This derivatization causes the decline of the charge density around the $^{m}C{\bullet}G$ base pair, following with an obvious current reduction. Owing to the velocity of the bromine derivatization is faster than that of the methylation; the general signal can promptly reflect the methylation status of the DNA. By continuously measuring the current decrease ratio, the monitoring of the dynamic process of DNA methylation can be achieved. In this chapter, we describe in detail the protocols of this method, including the label of DNA probe with electrochemical tags, the construction of the long-range electron transfer pathway on electrochemical sensor, the operation of the DNA methylation and bromine derivatization, as well as the continuous measurements of the conversed signals. This method may be potential for the application in biological research, disease diagnostics, and other areas.

Keywords DNA methylation, Dynamic process, Bromine derivatization, Long-range electron transfer, Continuous electrochemical determination

1 Introduction

DNA methylation exists in all living creatures [1]. In prokaryotes, it mainly occurs on the cytosine (C) in CCA/TGG and GATC, whereas in eukaryotes, it mostly occurs on the C, purine, and guanine [2]. DNA methylation is mostly linked to the activity of

Bi-Feng Yuan (ed.), *DNA Modification Detection Methods*, Springer Protocols Handbooks,
https://doi.org/10.1007/978-1-0716-1229-3_6, © Springer Science+Business Media, LLC, part of Springer Nature 2022

DNA methyltransferases (MTases), which catalyze the selective transfer of methyl group on *S*-adenosyl methionine (SAM) to the fifth carbon of the C, resulting in a product of 5-methylcytosine (5-^{m}C) [3]. As an important mode of epigenetic modification, DNA methylation plays key roles in gene-expression control, cellular activity, as well as growth and aging processes [4–6]. Besides, the abnormal DNA methylation behavior often causes aberration in life activity. Therefore, the information of DNA methylation has been recognized as the hallmark of many diseases, such as gene defect, growth delay, and miscellaneous cancers [7–9], and it is of great significance to establish methodologies for the determination of DNA methylation.

To date, numerous methods have been proposed for the determination of DNA methylation, including bisulfite treatment-, biological recognition- and chemical interaction-based strategies [10]. Although sensitive and accurate determinations of methylated DNA can be achieved [11–16], these methods can merely provide static "snapshot" views of methylation status from each test. Thus, tedious repeated measurements are inevitable to obtain dynamic information on DNA methylation. Multiple measurements carried in different batches are laborious, time/reagent consuming, and most importantly, causing detection deviation. By contrast, continuous monitoring of the dynamic process of DNA methylation is an efficient way to obtain more objective and precise information, but unfortunately the relevant methodologies are rare.

The challenges of tracking the dynamic process of DNA methylation mainly include two aspects: timely sensing the minute changes and synchronously amplifying the target signals. Conventional methodologies like quartz crystal microbalance or surface plasmon resonance are capable of detecting tiny changes in real time but not sensitive enough, whereas polymerase chain reaction-based or bio-affinity-based techniques, although sensitive, are usually discontinuous [17–20]. Double-strand DNA owns π-stack around the phosphoric acid skeleton, which provides a capacity of long-term electron transfer for the DNA, and the efficiency of the electron transfer is greatly affected by the variations on the C•G base pair or the π-stack, such as base mismatch or deviation, DNA melting, macromolecule binding, or adsorption [21–24]. Accordingly, we propose a novel electrochemical strategy for continuous monitoring of dynamic DNA methylation process. A specific chemical modification based on $NaIO_4$/LiBr is employed in our strategy. Once the targeting methylation site of the DNA is methylated, the bromine derivatization will be occurred immediately [25], which causes a decline of the charge density around the methylated site, and an obvious decrease of the electrochemical signal. Because the velocity of the bromine derivatization is faster than that of the methylation, the general signal can promptly reflect the methylation status of the DNA. Meanwhile, no additional interactions or operations are required for this method, which ensures the continuity of the measurements.

This chapter aims to describe this electrochemical method for the continuous monitoring of DNA methylation process, which mainly includes the preparation of electrochemical labels, the construction of double-strand DNA on electrochemical sensor, the operation of DNA methylation and bromine derivatization, as well as the measurement of signal conversion. In comparison with other methods, this method can facilely monitor dynamic DNA methylation process over time. This workflow is not only helpful for the deep investigation of DNA methylation mechanism and the assessment of MTases activity, but also applicable for the early diagnostic of diseases and the screening of clinical drugs.

2 Materials

2.1 Reagents

Gold nanoparticle (1000 ppm, diameters: 5–10 nm) solution (Aladdin Co. Ltd.).

6-Ferrocenylhexanethiol (Aladdin Co. Ltd.).

6-Mercapto-1-hexanol (Aladdin Co. Ltd.).

DNAs (TaKaRa Co. Ltd.), the sequences of the DNAs are as below:

Probe DNA: SH-C6-5′-CTCATGTACCGGACTAGTCC-3′-C6-SH.

Complementary DNA: 5′-GGACTAGTCCGGTACATGAG-3′.

Deerskin and Al_2O_3 powders with the diameters of 0.5, 0.3, 0.05 μm.

DNMT3a protein and SAM (New England BioLabs Inc.).

LiBr and $NaIO_4$ (Aladdin Co. Ltd.).

2.2 Solutions

Tris(2-carboxyethyl)phosphine (TCEP) solution (500 mM Aladdin Co. Ltd.).

Tris–HCl buffer (10 mM, pH 7.9): 10 mM Tris, 6.4 mM HCl.

TE buffer (10 mM, pH 7.9): 10 mM Tris–HCl buffer, 1 mM EDTA.

10× NE buffer 2 (500 mM, pH 7.9): 500 mM NaCl, 100 mM Tris–HCl, 100 mM $MgCl_2$, 10 mM DTT (New England BioLabs Inc.).

PBS buffer (10 mM, pH 7.4): 1.9 mM KH_2PO_4, 8.1 mM Na_2HPO_4, 10 mM NaCl.

DEPC water (Sangon Biotech. Co. Ltd.).

Ethyl alcohol (50%).

H_2SO_4 (50 mM).

Piranha solution (30% H_2O_2: 98% H_2SO_4 = 3:7, v:v).

5 mM $K_3[Fe(CN)_6]$

100 mM KCl.

2.3 Equipment

Centrifugal machine (COVARIS Inc.).

Shake device (COVARIS Inc.).

Pipettes with capacity of 100, 50, 10 μL (COVARIS Inc.).

Ultrasonic device (COVARIS Inc.).

Electrochemical workstation with three-electrode system (CHI 1230A, CH Instruments Ins.). The three-electrode system includes gold working electrode, Ag/AgCl reference electrode, and platinum wire counter electrode.

Nitrogen cylinder.

Thermostat water bath.

Magnetic heating agitator with a 0.5 cm stirrer (COVARIS Inc.).

Micro tubes with capacity of 5, 1, 0.5, 0.2 mL.

3 Methods

The entire procedure from the electrochemical label preparing to the continuous measurement of dynamic methylation process takes approximately 17 h of elapsed time, and it is visually summarized in Figs. 1, 2, and 3.

1. Electrochemical label preparation (6 h of elapsed time).
2. Hybridization of DNA strands (2 h of elapsed time).
3. Pretreatment of the gold electrode (1 h of elapsed time).
4. Assembling of DNA strands, 6-mercapto-1-hexanol (MCH), and the polymeride constructed in **step 1** above (9 h of elapsed time).
5. DNA methylation and bromine derivatization (2 h of elapsed time).
6. Continuous measurements (synchronization with **step 5** above).

The detailed information on the electrochemical label preparation (**step 1**) is described in Subheading 3.1, followed by detailed description of the electrochemical sensor construction (**steps 2–4**), the DNA methylation and bromine derivatization, as well as the continuous measurements in Subheadings 3.2–3.4, respectively. It is worth noticing that **steps 2** and **3** can be performed at the same time with **step 1** owing to the simple operation procedures.

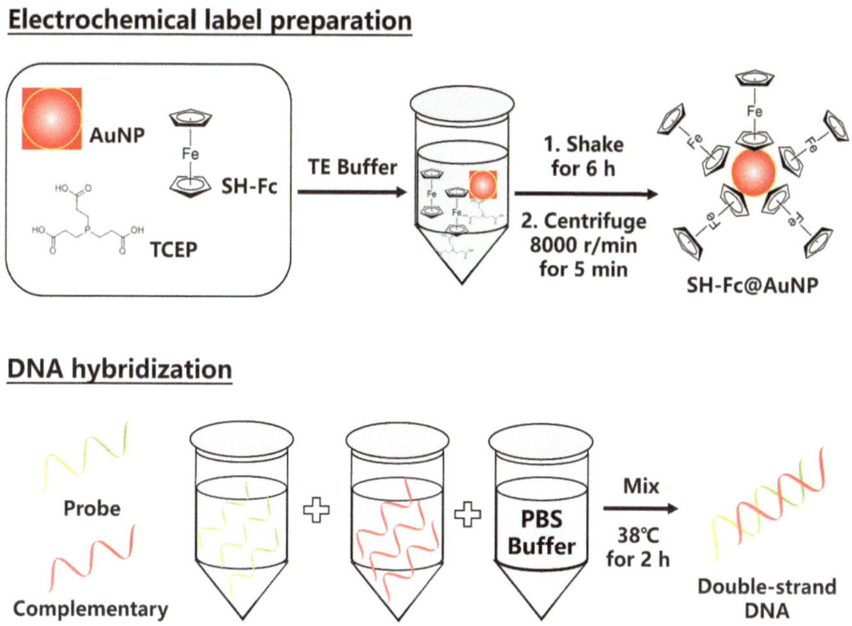

Fig. 1 Electrochemical label (SH-Fc@AuNPs) preparation and DNA hybridization

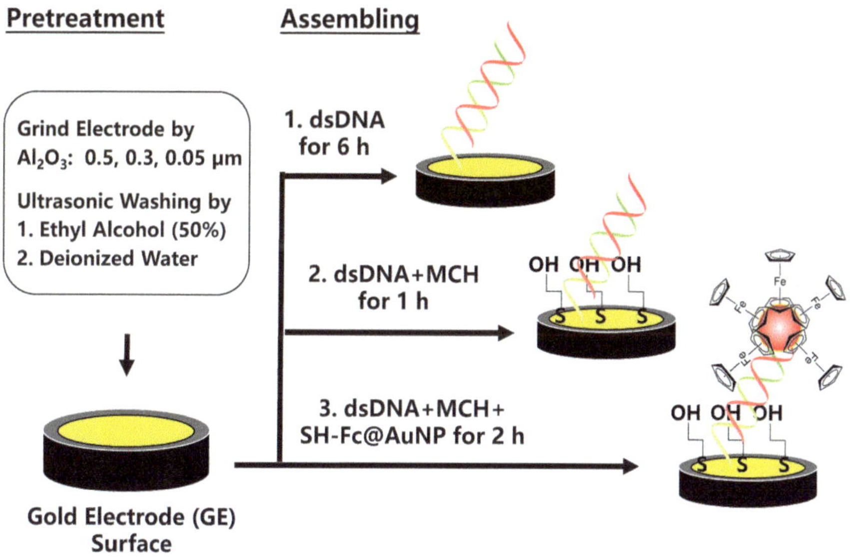

Fig. 2 Pretreatment of gold electrode and electrode assembling

3.1 Electrochemical Label Preparation

The electrochemical label is a polymeric material made of the gold nanoparticle (AuNP) and 6-ferrocenylhexanethiol (SH-Fc). Undergo the interaction of Au–S bond, large amount of SH-Fc is modified on the AuNP. In order to improve the efficiency, TCEP is employed to activate the sulfhydryl.

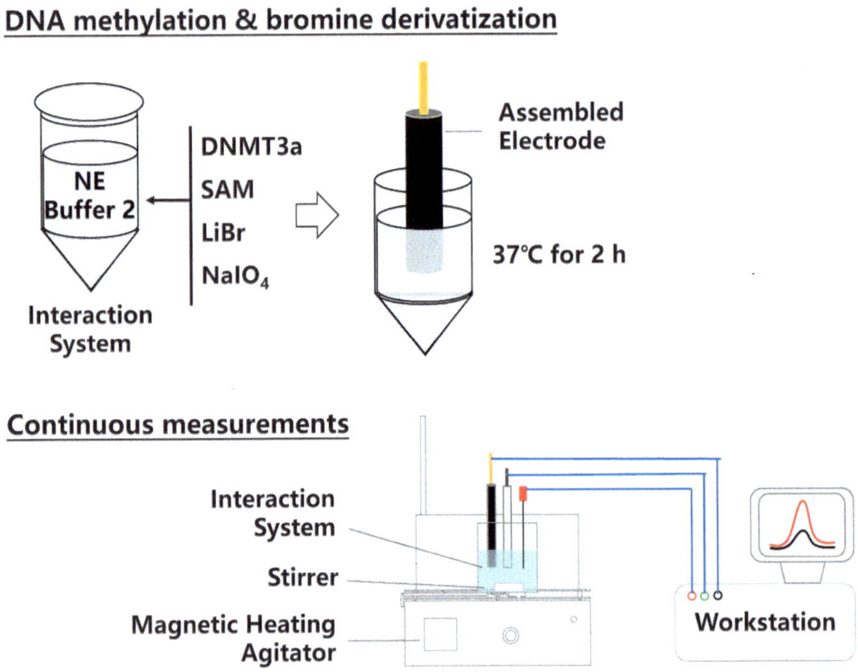

Fig. 3 DNA methylation and bromine derivatization and electrochemical continuous measurement

1. Mix 0.5 mL of 500 mM TCEP solution and 25 mL TE buffer.
2. Weigh 100 mg of SH-Fc and dissolve it using the solution prepared above.
3. Slowly drip 0.5 mL of 1000 ppm AuNP solution.
4. Completely shake up the solution and split charging it into micro tubes (5 mL per tube).
5. Put the micro tubes onto shake device, shaking for 6 h under low speed.
6. Centrifuge them at 10,000 r/min for 3 min and then discard the supernatant in each micro tube.
7. Add 5 mL TE buffer into each micro tube and shake fully to disperse the product at the bottom.
8. The result solutions (SH-Fc@AuNPs) are rufous, store them in refrigerator at 4 °C for further use (*see* **Note 1**).

3.2 Electrochemical Sensor Construction

The electrochemical sensor consisted of three parts, including the gold electrode, the double-strand DNA (dsDNA) modified on the electrode surface, and the prepared electrochemical label assembled on the top of the dsDNA.

1. Prepare 500 μL of 10 μM probe DNA and 20 μM complementary DNA solutions with TE buffer.

2. Mix 10 μL of 10 μM probe DNA, 10 μL of 20 μM complementary DNA, 10 μL of the PBS solution, and 36 μL of the deionized water.
3. Incubate the mixture at 38 °C for 2 h to hybridize the DNA.

3.3 Pause Point: Pretreatment of the Gold Electrode

1. Steep the gold electrode into 15 mL of *Piranha* solution for 15 min.
2. Wash the electrode with deionized water.
3. Add some Al_2O_3 powder (diameter: 0.5 μm) on the surface of deerskin and wet it with deionized water.
4. Grind the gold electrode on the deerskin with the Al_2O_3 powder and wash the electrode with deionized water.
5. Repeat **steps 3 and 4** using the Al_2O_3 powder with the diameters of 0.3 and 0.05 μm orderly.
6. Prepare 15 mL of ethyl alcohol (50%) in the breaker and place the electrode in the breaker to perform ultrasonic washing for 3 min twice, then dry under nitrogen.
7. Prepare 10 mL of 50 mM H_2SO_4 in the electrolytic tank, and use the polished gold electrode to perform cyclic scanning with Ag/AgCl reference electrode and platinum wire counter electrode (scan range: −0.2 to 1.6 V, scan rate: 0.05 V, scan cycle: 10) for the electrode activation.
8. Wash the electrode with deionized water and dry the surface under nitrogen.
9. Prepare 10 mL of 5 mM $K_3[Fe(CN)_6]$ solution which contained 100 mM KCl in the electrolytic tank, and use the activated gold electrode to perform CV tests with Ag/AgCl reference electrode and platinum wire counter electrode (scan range: −0.1 to 0.6 V, scan rate: 0.05 V, scan cycle: 1) to assess the polishing status.
10. When the potential difference between the oxidation–reduction signal peak in the CV graph is <80 mV, the gold electrode polishing is achieved (*see* **Note 2**).
11. Wash the electrode with deionized water and store it in ethyl alcohol (50%) at room temperature for further use.

3.3.1 Pause Point: Store at −20 °C for Further Use or Use Immediately

1. Wash the polished gold electrode with deionized water and dry under nitrogen.
2. Mix 0.5 mL of 500 mM TCEP solution with 49.5 mL TE buffer.
3. Mix 60 μL of the dsDNA and 6 μL of the 5 mM TCEP solution in micro tube.

4. Drip mixture onto the surface of the polished gold electrode, then reacts at 4 °C for 6 h to link the hybridized DNA to the electrode surface.
5. Wash the electrode with deionized water and dry under nitrogen.
6. Dilute the MCH to 0.5 mM by the TE buffer and drip 50 μL of it onto the electrode surface, then react at 4 °C for 1 h to block the electrode surface.
7. Wash the electrode with deionized water and dry under nitrogen.
8. Drip 100 μL of the prepared SH-Fc@AuNPs onto the electrode surface, then react at 4 °C for 2 h to modify the SH-Fc@AuNPs onto DNA.
9. Wash the electrode with deionized water and dry under nitrogen.
10. Take 15 mL of the PBS buffer into the electrolytic tank, use the electrode to perform CV tests with Ag/AgCl reference electrode and platinum wire counter electrode (scan range: −0.1 to 0.6 V, scan rate: 0.05 V, scan cycle: 1) to assess the assembling status.
11. Once a cyclic oxidation–reduction signal peak of $Fe^{3+/2+}$ presences in the CV graph, the assembling of dsDNA and electrochemical label is achieved (*see* **Note 3**).

3.4 DNA Methylation and Bromine Derivatization

DNA methylation requires the catalysis of transmethylase and the substrate support of SAM. DNMT3a protein is a de novo methylase, it can methylate normal DNA without the assist of methylated site. Another specific interaction is the bromine derivatization, which relates to the oxidation of $NaIO_4$; a bromine ion of LiBr can transfer to the C4 site of the cytosine once it was methylated. Owing to the little influence of the chemical micromolecule on the activity of DNMT3a protein, it provides a potential of cascade reaction in one-pot way. This section mainly focuses on the establishment of the interaction system of DNA methylation and bromine derivatization.

1. Mix 300 μL of 1 μM DNMT3a protein, 300 μL of 1 mM SAM, 300 μL of 8 mM $NaIO_4$, 300 μL of 32 mM LiBr, 300 μL of 10× NE Buffer 2, and 1.5 mL of the PBS buffer (SAM, $NaIO_4$, and LiBr solutions are in DEPC water, final concentrations: 100 nM DNMT3a protein, 100 μM SAM, 0.8 mM $NaIO_4$, and 3.2 mM LiBr) in the micro glass tank.
2. Incubate the mixture solution at 37 °C, and then insert the modified gold electrode as described above. The DNA methylation and bromine derivatization occurred; the total time of the reaction is ~100 min.

3.5 Continuous Measurements

To measure the dynamic DNA methylation process, a continuous electrochemical assay is performed during the DNA methylation. The methylation and bromine derivatization cause an obvious decline of the electron density around the methylated site, and this status is reflected by a decrease of the electrochemical signal. By measuring the electrochemical signal decline in different methylation phases, the methylation process can be investigated. To avoid the influence of DNMT3a protein on the DNA strands during the electrochemical assays, a magnetic heating agitator is employed to clean the adsorbed protein. This section mainly focuses on the description of the continuous measurement procedure.

1. Followed by the **step 2** in Subheading 3.3, place a stirrer with the length of 0.5 cm into the solution (showed as Fig. 1f).
2. Use the modified gold electrode to perform DPV test in the mixture with Ag/AgCl reference electrode and platinum wire counter electrode (scan range: −0.1 to 0.6 V, potential step: 4 mV, amplitude: 50 mV, pulse width: 0.05 s, pulse cycle: 0.2 s).
3. Record the test result as I_0.
4. Incubate at 37 °C for 5 min.
5. Start the magnetic stirring at 800 r/min for 1 min (*see* **Note 4**).
6. Continue the DPV test with the same condition and record the result as I_1.
7. Repeat the **steps 4–6** and record the results as I_2, I_3, I_4 ... I_{20}.
8. Calculate the current decline ratio ($1 - I/I_0$) of the signals after each reaction period and draw the conversion curve which reflects the generative process of the methylated DNA.

4 Notes

1. Electrochemical label preparation.

   After the addition of the TE buffer, ensure to shake the solution completely.

   If the result solution is in dusty blue, it indicated that the AuNPs are aggregated, which needs to disperse again by ultrasonication.

   Salinity is an important influence factor. High salinity leads to the aggregation of the AuNPs. Ensure to use low salinity buffer to perform the experiments.

   Diameter is another influence factor; ensure to use the AuNPs with the diameter of 5–10 nm. If the diameter is larger than 15 nm, the centrifugal rotational speed should adjust to 8000 r/min.

2. Pretreatment of the gold electrode.
   - If the potential difference of the oxidation–reduction signal peak is above 80 mV, it can be caused by one of five issues:
     - The electrode is auld.
     - The grind way is incorrect.
     - The electrode surface exists as residual Al_2O_3 powder.
     - The electrode activation is incomplete.
     - The KCl is absent in the $K_3[Fe(CN)_6]$ solution.

     Check the electrode surface and avoid the ones that have scratches or corrosion. Keep the electrode in vertical status during the grind and keep the grind route like "8." Ultrasonic washing is required after the grind with Al_2O_3, and ensure the electrode surface is clean. The number of the cyclic scanning in H_2SO_4 should be strictly controlled; more or less cycles will influence the electrode property.
3. Assembling of dsDNA and electrochemical label.
   - The assembled electrode is not suitable to store for a long time owing to the timeliness of the dsDNA. It prefers to be used after the electrode is ready.
   - To ensure the assembling is successful, the impedance tests can be performed after each assembling stage. In theory, the impedance of the electrode is continuously enhancing when assembling dsDNA and MCH, and then reduces slightly when modifying the SH-Fc@AuNPs owing to the establishment of the electron transfer approach. In this way, the assembling procedure can be assessed.
4. Continuous measurements.
   - The DPV test must be paused during the magnetic stirring.
   - Ensure the measured mixture is steady before the test.
   - Avoid the stirrer touching the electrode in the mixture when it is rotating.
   - If the stirrer is larger than 0.5 cm, or the turbulent flow is violent, the speed of the rotation should be reduced, and the stirring time can be extended.

Acknowledgments

This work was supported by the National Natural Science Foundations of China (21422510, 21375154, and 21675180), the Scientific Technology Project of Guangdong Province (2016B010108007, 2014A040401022, 2015A030401033, and 2017B020221001), and the Scientific Technology Project of Guangzhou City (201604020145).

References

1. Dahl C, Guldberg P (2003) DNA methylation analysis techniques. Biogerontology 4:233–250
2. Tucker KL (2001) Methylated cytosine and the brain: a new base for neuroscience. Neuron 30:649–652
3. Ehrlich M, Gama MA, Huang LH, Midgett RM, Kuo KC, McCune RA et al (1982) Amount and distribution of 5-methylcytosine in human DNA from different types of tissues or cells. Nucleic Acids Res 10:2709–2721
4. Cottrell SE (2004) Molecular diagnostic applications of DNA methylation technology. Clin Biochem 37:595–604
5. Gertz J, Varley KE, Reddy TE, Bowling KM, Pauli F, Parker SL et al (2011) Analysis of DNA methylation in a three-generation family reveals widespread genetic influence on epigenetic regulation. PLoS Genet 7:E1002228
6. Jones PA, Laird PW (1999) Cancer epigenetics comes of age. Nat Genet 21:163–167
7. Trasler JM (2006) Gamete imprinting: setting epigenetic patterns for the next generation. Reprod Fertil Dev 18:63–69
8. Esteller M, Corn PG, Baylin SB, Herman JG (2001) A gene hypermethylation profile of human cancer. Cancer Res 61:3225–3229
9. Laird PW (2003) The power and the promise of DNA methylation markers. Nat Rev Cancer 3:253–266
10. Zhang L, Xu YZ, Xiao XF, Chen J, Zhou XQ et al (2015) Development of techniques for DNA-methylation analysis. Trends Anal Chem 72:114–122
11. Gonzalgo ML, Jones PA (1997) Rapid quantitation of methylation differences at specific sites using methylation-sensitive single nucleotide primer extension (Ms-SNuPE). Nucleic Acids Res 25:2529–2531
12. Aggerholm A, Guldberg P, Hokland M, Hokland P (1999) Extensive intra and inter individual heterogeneity of pl5INK4B methylation in acute myeloid leukemia. Cancer Res 59:436–441
13. Fraga MF, Uriol E, Borja LD, Berdasco M, Esteller M, Cañal MJ et al (2002) High-performance capillary electrophoretic method for the quantification of 5-methyl 2-deoxycytidineingenomic DNA: application to plant, animal and human carcinoma tissues. Electrophoresis 23:1677–1681
14. Rodríguez-Lopez CM, Guzmán AB, Lloyd AJ, Wilkinson MJ (2010) Direct detection and quantification of methylation in nucleic acid sequences using high resolution melting analysis. Anal Chem 82:9100–9108
15. Zhang J, Song J, Nie C, Liu L, Lv F, Wang S et al (2014) Associated analysis of DNA methylation for cancer detection using CCP-based FRET technique. Anal Chem 86:346–350
16. Xu YZ, Niu C, Xiao XF, Zhu WY, Dai Z, Zou XY (2015) Chemical-oxidation cleavage riggered isothermal exponential amplification reaction for attomole genespecific methylation analysis. Anal Chem 87:2845–2851
17. Daems D, Knez K, Delport F, Spasica D, Lammertyn J (2016) Real-time PCR melting analysis with fiber optic SPR enables multiplex DNA identification of bacteria. Analyst 141:1906–1911
18. Huang W, Qi CB, Lv SW, Xie M, Feng YQ, Huang WH, Yuan BF (2016) Determination of DNA and RNA methylation in circulating tumor cells by mass spectrometry. Anal Chem 88:1378–1384
19. Meissner A, Gnirke A, Bell GW, Ramsahoye B, Lander ES, Jaenisch R (2005) Reduced representation bisulfite sequencing for comparative high-resolution DNA methylation analysis. Nucleic Acids Res 33:5868–5877
20. Wang CH, Lai HC, Liou TM, Hsu KF, Chou CY, Lee GB (2013) A DNA methylation assay for detection of ovarian cancer cells using a HpaII/MspI digestion- based PCR assay in an integrated microfluidic system. Microfluid Nanofluid 15:575–585
21. Kelley OS, Barton KJ (1999) Electron transfer between bases in double helical DNA. Science 283:375–381
22. Lewis DF, Wu TF, Zhang YF, Letsinger LR, Greenfield RS, Wasielewski RM (1997) Distance-dependent electron transfer in DNA hairpins. Science 277:673–667
23. Mohamadi M, Mostafavi A, Torkzadeh-Mahani M (2015) Electrochemical determination of biophenol oleuropein using a simple label-free DNA biosensor. Bioelectrochemistry 101:52–57
24. Claesson MJ, Cusack S, O'Sullivan O, Greene-Diniz R, Weerd HD, Flannery E et al (2016) Highly specific SNP detection using 2D graphene electronics and DNA strand displacement. Proc Natl Acad Sci U S A 113:7088–7093
25. Bareyt S, Carell T (2008) Selective detection of 5-methylcytosine sites in DNA. Angew Chem Int Ed 47:181–184

Chapter 7

Electrogenerated Chemiluminescence Method for Determination of 5-Hydroxymethylcytosine in DNA

Shangxian Ma, Yan Li, and Honglan Qi

Abstract

5-methylcytosine (5-mC) is regarded as the fifth base of DNA, while 5-hydroxymethylcytosine (5-hmC) is regarded as the sixth base of DNA. With the catalysis of ten–eleven translocation proteins, 5-mC can be converted to 5-hmC, resulting in DNA hydroxymethylation. Therefore, the discrimination and determination of 5-hmC in DNA are getting increasing interests in the field of epigenetics. Here, we introduce a selective and sensitive method for the determination of 5-hmC in DNA by incorporating the specific oxidation of 5-hmC to 5-formylcytosine (5-fC) by $KRuO_4$ and electrogenerated chemiluminescence method. The detection limit was 1.4×10^{-13} M for 5-hmC containing DNA. The combination of selective 5-hmC oxidation and sensitive ECL method is a promising approach for the determination of 5-hmC in DNA.

Keywords Electrogenerated chemiluminescence, 5-Hydroxymethylcytosine, DNA, $KRuO_4$, Oxidation

1 Introduction

There are four canonical DNA nucleosides dA, dC, dG, and dT [1]. 5-Methylcytosine (5-mC) is regarded as the fifth base of DNA. 5-hydroxymethylcytosine (5-hmC) is regarded as the sixth base of DNA [2]. With the catalysis of ten–eleven translocation (TET) proteins, 5-mC can be converted to 5-hmC, resulting in DNA hydroxymethylation. 5-hmC is abundant in brain and embryonic stem cells and the distribution of 5-hmC is different in different tissues of human body. For example, the level of 5-hmC is low in the tumors of heart, lungs, spleen, and breast, while the level of 5-hmC is high in the brain. 5-hmC can be considered as a biomarker for disease diagnosis, treatment, and prognosis. The level of 5-hmC is much lower than that of 5-mC. Therefore, discrimination and determination of 5-hmC in DNA have gradually getting increasing interests in the fields of epigenetics with selectivity and sensitivity.

Bi-Feng Yuan (ed.), *DNA Modification Detection Methods*, Springer Protocols Handbooks,
https://doi.org/10.1007/978-1-0716-1229-3_7,

A variety of methods have been established for the determination of 5-hmC in DNA, such as liquid chromatography-mass spectrometry (LC-MS) [3], thin layer chromatography (TLC) [4], single-molecule real-time (SMRT) sequencing [5], fluorescence [6], electrochemical [7], and electrogenerated chemiluminescence (ECL) methods [8]. Each method has its own advantages and disadvantages. For example, SMRT needs the enrichment of 5-hmC containing DNA and additional fluorescent tags. TLC requires radioactive substrates, which are limited with unhealthy radioactive waste. LC-MS is a powerful method for 5-hmC in DNA, but the instruments are expensive and sophisticated.

From the recognition mechanism, there are usually two types of approaches for the determination of 5-hmC in DNA. One is the use of anti-5-hmC antibody as the 5-hmC recognition element for the enrichment and determination of 5-hmC containing DNA with high selectivity. However, it does not have single-base resolution and has some limitations, such as production from limited shelf life, living animals in vivo, easy denaturation, and high cost. The other alternative approach for identifying 5-hmC is chemical modification of 5-hmC. One popular approach is the employment of T4 β-glucosyltransferase for glycosylation reaction, in which the 5-hmC can be glycosylated by using T4 β-glucosyltransferase, but the 5-mC and cytosine cannot be glycosylated by T4 β-glucosyltransferase. Therefore, the glucose group on the 5-hmC in DNA can be further modified or labeled or restricted with endonuclease digestion for analysis. For example, He's group reported that 5-hmC was firstly glycosylated with an azide-substituted glucose by β-glucosyltransferase and then modified by a biotin tag through click chemistry [9], which providing a selective chemical labeling method for 5-hmC. Natural enzyme-based methods suffer some intrinsic drawbacks such as susceptibility to enzyme denaturation and low stability. Another way is the employment of the $KRuO_4$ oxidation, firstly reported by Balasubramanian and coworkers [10]. They found that $KRuO_4$ could exclusively oxidize 5-hmC to form 5-formylcytosine (5-fC) with high yield. After that, various detection methods incorporating with $KRuO_4$ oxidation have been developed for the determination of 5-hmC in DNA. For instance, Zhou's group reported a fluorescence method for the determination of 5-hmC in DNA by incorporation of $KRuO_4$ oxidation with fluorescence detection via a cationic conjugated polymer [11].

Electrogenerated chemiluminescence is a kind of luminescence occurred at/near the surface of electrode as a result of electrochemical reactions and chemiluminescence reactions [12]. ECL is superior to photoluminescence, chemiluminescence, and electrochemistry in some aspects, which has extremely low detection limits and high sensitivity. ECL method has been getting increasing interests in clinical diagnosis and pharmaceutical

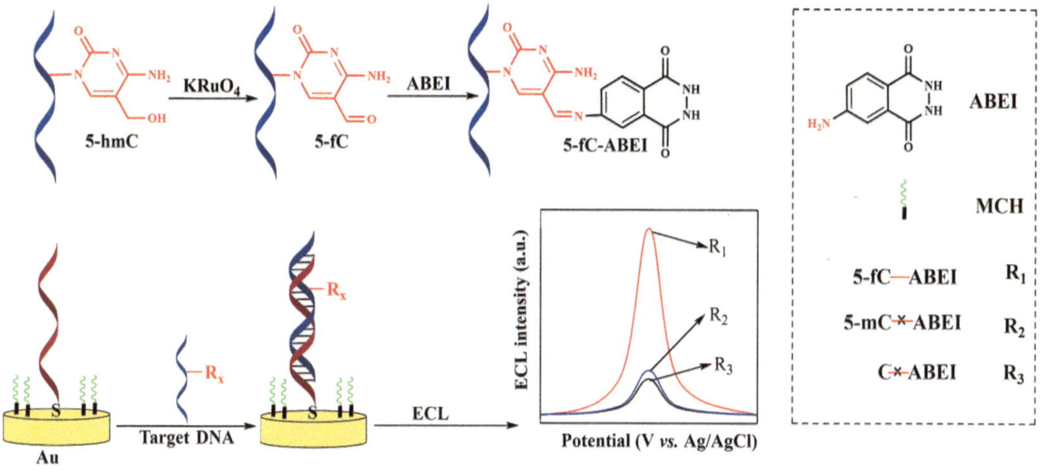

Fig. 1 Schematic diagram of the ECL method for the determination of 5-hmC containing DNA. A thiolated capture probe (ss DNA, 35-mer) for the target DNA containing 5-hmC was self-assembled on a gold surface. The 5-hmC in the target DNA was selectively transformed to 5-fC via oxidation by $KRuO_4$ and then subsequently labeled with ABEI. The ABEI-labeled target DNA was hybridized with the capture probe on the electrode, resulting in a strong ECL emission. (Reproduced from ref. [8] with permission from American Chemical Society)

analysis, etc. [13]. Our group have developed several ECL biosensing methods to discriminate and determine 5-hmC containing DNA based on 5-hmC antibody recognition [14], $KRuO_4$ oxidization [8, 15], T4 β-glucosyltransferase glycosylation [16, 17], and M.HhaI methyltransferase coupled with cysteamine modification [18].

This chapter provides the detailed protocol for determination of 5-hmC containing DNA by incorporating N-(4-aminobutyl)-N-ethylisoluminol (ABEI)-based electrogenerated chemiluminescence assay and the $KRuO_4$ oxidation of 5-hmC (Fig. 1) [8]. N-(4-ainobutyl)-N-ethylisoluminol (ABEI), as an ECL emitter and label, is a derivative of luminol. ABEI can be electrochemically oxidized in the presence of hydrogen peroxide to form excited-state species that emit light [19]. This method is selective and sensitive for 5-hmC containing DNA with a detection limit of 1.4×10^{-13} M and can be used to detect 5-hmC containing DNA in biological samples.

2 Materials

All materials are listed in order of time of use for the corresponding part of the protocol listed below.

2.1 Solutions

1. 0.5 M disodium EDTA: 9.306 g of $EDTA\text{-}Na_2{\cdot}2H_2O$ was dissolved in 50 mL H_2O.

2. 3 M sodium acetate (NaAc): 10.202 g of $CH_3COONa{\cdot}3H_2O$ was dissolved in 25 mL H_2O.
3. 0.1 M PB buffer (pH 7.4): 1.482 g of $NaH_2PO_4{\cdot}2H_2O$ and 14.508 g of $Na_2HPO_4{\cdot}12H_2O$ are dissolved in 497 mL H_2O and adjust to pH 7.4 with NaOH or HCl and fixed the volume to 500 mL with H_2O.
4. N-(4-Aminobutyl)-N-ethylisoluminol (ABEI): 5 mg ABEI was dissolved in 1 mL 0.1 M HCl and then fixed the volume to 100 mL with H_2O as storage solution. 1.0×10^{-6} M ABEI solution was prepared by diluting the ABEI storage solution with 0.1 M PB buffer (pH 7.4).
5. Saturated $KRuO_4$: dissolve $KRuO_4$ with H_2O.
6. 1 mM 6-mercapto-1-hexanol (MCH): 14 μL of 7 M MCH was added in 986 μL ethanol to obtain 0.1 M MCH as storage solution. Then 10 μL of 0.1 M MCH was added in 990 μL of 10 mM Tris–HCl buffer (pH 7.4).
7. 0.1 M Tris: 0.606 g trizma base was dissolved in 50 mL H_2O.
8. 10 mM Tris–HCl buffer (pH 7.4): 50 mL of 0.1 M Tris was mixed with 42 mL of 0.1 M HCl, and fixed the volume to 500 mL with H_2O. The pH of the solution is adjusted with HCl at 7.4.
9. 1 M Tris–HCl buffer (pH 8.0): 6.060 g of trizma base was dissolved in 40 mL H_2O. The pH of the solution is adjusted with HCl at 8.0 and fixed the volume to 50 mL with H_2O.
10. 10 mM Tris–HCl buffer containing 1 mM disodium EDTA (pH 8.0): 5 mL of 1 M Tris–HCl buffer (pH 8.0) was mixed with 1 mL of 0.5 M disodium EDTA and fixed the volume to 500 mL with H_2O.
11. 0.1 M Tris–HCl buffer containing 7 mM TCEP (pH 8.5): 0.606 g of trizma base, 0.100 g of TCEP were dissolved in 45 mL H_2O. The pH of the solution is adjusted with HCl at 8.5 and fixed the volume to 50 mL with H_2O.
12. 10 mM Tris–HCl buffer containing 1 mM disodium EDTA and 1.0 M NaCl (pH 7.4): 0.186 g of EDTA-$Na_2{\cdot}2H_2O$, 29.22 g of NaCl are dissolved in 50 mL of 0.1 M Tris and fixed the volume to 500 mL with H_2O. The pH of the solution is adjusted with HCl at 7.4.
13. 0.1 M PBS buffer (pH 7.4): 1.482 g of $NaH_2PO_4{\cdot}2H_2O$, 14.508 g of $Na_2HPO_4{\cdot}12H_2O$, and 2.922 g of NaCl are dissolved in 497 mL H_2O and the pH of the solution was adjusted to pH 7.4 with NaOH or HCl and fixed the volume to 500 mL with H_2O.

2.2 Equipment

ECL measurements were performed on an MPI-A ECL detector (Xi'an Remax Analysis Instruments Co. Ltd. China) (*see* **Note 1**). Electrochemical experiments were performed on a CHI 660 electrochemical workstation (Shanghai Chenhua Instruments Co. Ltd. China). The UV–Vis absorption spectrum was recorded on a UV–Vis spectrophotometer (UV-2450, Shimadzu Corporation, Japan). The polyacrylamide gel electrophoresis was performed on the Bio-Rad electrophoresis analyzer (China).

2.3 DNA Sequence

The sequences of DNA used in this work are as follows (*see* **Note 2**):

1. DNA probe S1: 5′-HS-$(CH_2)_6$-TGC GTG CGC GCT CCC GAG TCG ACC TCC GTA GTC TT-3′
2. Complementary DNA with one 5-hmC modification (DNA S2): 5′-AAG ACT ACG GAG GTC GAC $T^{5\text{-hm}}$CG GGA GCG CGC ACG CA-3′
3. Complementary DNA with three 5-hmC modification (DNA S3): 5′-AAG ACT $A^{5\text{-hm}}$CG GAG GTC GAC $T^{5\text{-hm}}$CG GGA GCG CGC $A^{5\text{-hm}}$CG CA-3′
4. Complementary DNA with 5-mC modification (DNA S4): 5′-AAG ACT ACG GAG GTC GAC $T^{5\text{-m}}$CG GGA GCG CGC ACG CA-3′
5. Complementary DNA with normal cytosine as control (DNA S5): 5′-AAG ACT ACG GAG GTC GAC TCG GGA GCG CGC ACG CA-3′

The five oligonucleotides were dissolved in 10 mM Tris–HCl buffer containing 1 mM EDTA (pH 8.0) to the desired concentrations and stored at −20 °C (*see* **Note 3**).

3 Methods

3.1 Oxidation of DNA by $KRuO_4$

10 μL of saturated $KRuO_4$ was mixed with 200 μL of 1 μM DNA S2. The reaction was incubated in ice-water bath for 1 h. Excess $KRuO_4$ was removed through dialysis with molecular weight cutoff membrane (1000 Da) against water to obtain 5-fC DNA (*see* **Note 4**).

3.2 ABEI Labeling of 5-fC DNA

40 μL of 100 μM ABEI was mixed with 200 μL the obtained 5-fC DNA and the reaction was incubated in 37 °C water bath for 12 h. Then 24 μL of 3 M sodium acetate and 600 μL ethanol were added into the mixture of ABEI and 5-fC DNA. The final mixture was centrifuged at 12,000 rpm (2-16R Ultrasonic, Hunan Hengnuo Instrument Co. Ltd., China) for 5 min, and the supernatant was removed. Then the precipitate was redissolved with 600 μL of 70% ethanol, centrifuged at 12,000 rpm (about 10000 g) for 5 min, and

the supernatant was removed. The precipitate was dried and redissolved in 10 mM Tris–HCl containing 1 mM EDTA (pH 8.0) to the desired concentrations and stored at −20 °C.

3.3 Characteristic of ABEI-Labeled 5-fC-Containing DNA

The obtained ABEI-labeled 5-fC-containing DNA can be characterized using UV–Vis absorption spectroscopy and PAGE analysis. The presence of a blue fluorescence emission band taken under 365 nm UV lamp is an evidence of the ABEI-labeled 5-fC-containing DNA. Two peaks at 284 and 240 nm corresponding to the absorption spectra of 5-fC-DNA and ABEI can be used as indicator from the UV–Vis absorption spectroscopy of the product solution.

3.4 Polyacrylamide Gel Electrophoresis (PAGE)

1. 30% polyacrylamide: 29 g acrylamide and 1 g methyl diacrylamide are dissolved in 100 mL H_2O.
2. 10× TBE (1 M Tris–HCl, 830 mM boric acid, 10 mM EDTA, pH 8.2): 30 g Trizma base, 0.930 g of EDTA-$Na_2{\cdot}2H_2O$, 12.82 g boric acid are dissolved in 250 mL H_2O and adjust to pH 8.2.
3. 1× TBE (100 mM Tris–HCl, 83 mM boric acid, 1 mM EDTA, pH 8.2): 3.0 g Trizma base, 0.093 g EDTA-$Na_2{\cdot}2H_2O$, 1.282 g boric acid are dissolved in 250 mL H_2O and adjust to pH 8.2.
4. 10% ammonium persulfate: 1 g ammonium persulfate is dissolved in 10 mL H_2O.
5. 20% polyacrylamide gel was prepared by mixing 10 mL of 30% polyacrylamide, 3 mL 10× TBE (1 M Tris–HCl, 830 mM boric acid, 10 mM EDTA, pH 8.2), 17 mL H_2O, 150 μL of 10% ammonium persulfate and 15 μL of tetramethylethylene diamine. Electrophoresis buffer was 1× TBE buffer. 2.5 μL DNA sample (1.0×10^{-6} M C-DNA, 5-mC-DNA, and 5-hmC-DNA-ABEI) was applied onto the lane of the PAGE gel, respectively; 150 V was applied on the lane for 120 min. Finally, the PAGE images were taken under daylight and 365 nm UV light.

3.5 Fabrication of ss-DNA-Modified Gold Electrode

1. The gold electrode was firstly polished with alumina powder with different sizes (such as 0.3 μm and 0.05 μm). Then the polished electrode was sonicated in acetone and water for several minutes. The obtained electrode was electrochemically treated in 0.1 M H_2SO_4 solution by consecutively cycling the potential in the range of 0 to +1.5 V with a scan rate of 0.5 V/s to get cyclic voltammogram characteristic of a clean Au electrode.
2. 100 μL of 100 μM DNA S1 in water was mixed with 100 μL of 0.1 M Tris–HCl containing 7 mM TCEP (pH 8.5) and incubated in the dark for 3 h at room temperature to reduce any

possible disulfide bonds and then diluted to 1 μM with 10 mM Tris–HCl buffer containing 1.0 mM EDTA and 1.0 M NaCl (pH 7.4) for immobilization (*see* **Note 5**).

3. The obtained gold electrode was immersed into 100 μL of 1 μM DNA S1 solution and incubated at 37 °C for several hours (such as 4 h). The DNA modified electrode was then washed twice with 10 mM Tris–HCl buffer (pH 7.4) to remove the nonspecifically adsorbed DNA S1. The modified electrode was then dipped into 100 μL of 1 mM MCH solution for 1 h. The resulting modified gold electrode was then thoroughly washed with 10 mM Tris–HCl buffer (pH 7.4) and used as the DNA S1-modified gold electrode.

3.6 Electrochemical Measurement

The modification of DNA S1 and the blocking of MCH can be confirmed using cyclic voltammetry and electrochemical impedance spectroscopy. Electrochemical measurement was done in 5 mM $[Fe(CN)_6]^{3-}$-5 mM $[Fe(CN)_6]^{4-}$-0.1 M KCl. The cyclic voltammogram (CV) parameters were set as following: Init $E = -0.2$ V; final $E = +0.6$ V; scan rate: 0.1 V/s. The electrochemical impedance spectroscopy (EIS) parameters were set as follows: Init $E = +0.284$ V, high frequency (Hz): 100,000; low frequency (Hz): 0.1; amplitude (V): 0.005. The decrease of peak current in the CVs and the increase of the charge transfer resistance in EIS can support the successful immobilization of DNA S1 on the surface of gold electrode.

3.7 ECL Measurement

The DNA S1 modified gold electrode was immersed in 100 μL of 10 mM Tris–HCl buffer containing 1 mM disodium EDTA and 1.0 M NaCl (pH 7.4) containing a fixed concentration of 5-fC-DNA-ABEI for 2 h at 37 °C, and then was washed with 10 mM Tris–HCl buffer (pH 7.4). A traditional three-electrode system was employed, including DNA S1 modified gold electrode as working electrode, platinum wire as counter electrode, and Ag/AgCl (saturated KCl) as reference electrode. The ECL measurement was performed using a triangular potential scan in the range of +0.2 to 0.8 V with the rate of 0.1 V/s in 0.1 M PBS (pH 7.4) containing 1 mM H_2O_2. The ECL parameters were set as following: Init $E = +0.2$ V; Final $E = +0.8$ V; scan rate: 0.1 V/s; sample rate (T/s): 10; amplifier Series: 3. The photomultiplier tube (PMT) was biased at −900 V (*see* **Note 1**). The concentration of 5-hmC containing DNA was quantified by an ECL intensity at +0.7 V (*see* **Note 6**). The ECL intensity was correlated with the concentration of 5-hmC containing DNA and could be used to discriminate 5-hmC containing DNA and 5-mC containing DNA. Slight changes in the ECL intensity were found for the other tested DNA sequences (Fig. 2). The ECL intensity at +0.7 V can be used to determine 5-hmC containing DNA with a low detection limit of

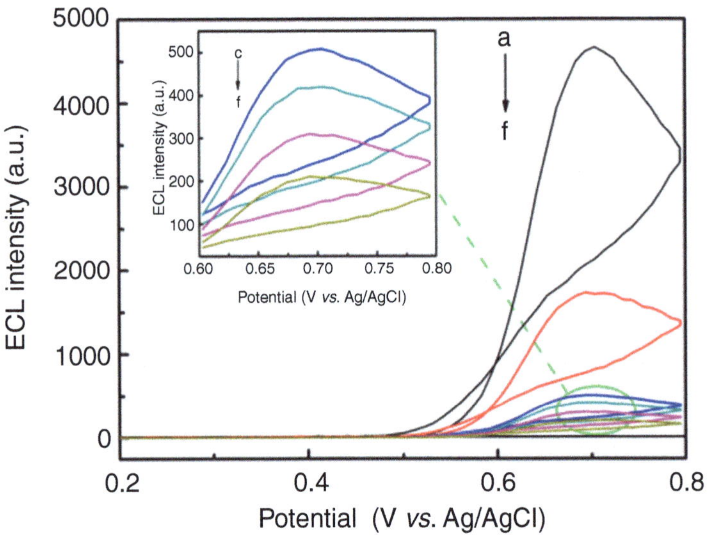

Fig. 2 ECL intensity versus potential profiles for different DNAs combined with ABEI; the concentration of all DNAs was 1.0×10^{-11} M: 5-hmC-containing DNA (DNA S3) after oxidization (line a), 5-hmC-DNA (DNA S2) after oxidization (line b), 5-hmC-DNA (DNA S2) without oxidization (line c), three-base mismatch DNA (DNA S6) after oxidization (line d), 5-mC-DNA (DNA S4) after oxidization (line e), noncomplementary DNA (DNA S7) after oxidization (line f), C-DNA (DNA S5) after oxidization (line g), and hybridization buffer without DNA (line h). Measurement conditions: 0.1 M PBS containing 1 mM H_2O_2 (pH 7.4); scan rate, 100 mV/s. (Reproduced from ref. [8] with permission from American Chemical Society)

1.4×10^{-13} M (Fig. 3). The proposed ECL method showed high sensitivity and selectivity for the determination of 5-hmC-containing DNA.

4 Notes

1. We use an MPI-A ECL detector (Xi'an Remax Electronics) for ECL measurement. Other instruments to measure ECL signal are also applicable. The ECL parameters can be variable according to the instruments.
2. The sequences of DNA used in this work were purchased from biological company and their base sequences are listed above. Different DNA sequence can be used according to the aim of work.
3. For better modification or self-assembling or hybridization of DNA, the DNA sample should be dissolved in a buffer solution. Different buffer solutions containing salt can be used according to the aim of the work.

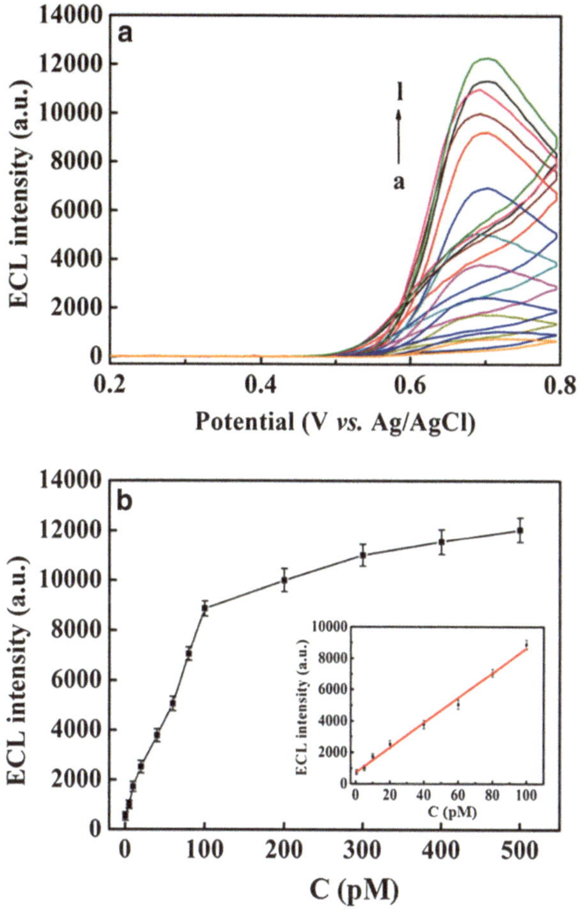

Fig. 3 (**a**) ECL responses to different concentrations of 5-hmC containing DNA (DNA S2) in 0.10 M PBS (pH 7.4) containing 1 mM H_2O_2. Line a-l, 5.0×10^{-13} M, 5.0×10^{-12} M, 1.0×10^{-11} M, 2.0×10^{-11} M, 4.0×10^{-11} M, 6.0×10^{-11} M, 8.0×10^{-11} M, 1.0×10^{-10} M, 2.0×10^{-10} M, 3.0×10^{-10} M, 4.0×10^{-10} M, and 5.0×10^{-10} M. (**b**) Relationship between the ECL intensity and DNA S2 concentration. Inset: calibration curve for DNA S2. (Reproduced from ref. [8] with permission from American Chemical Society)

4. The pore size of cutoff dialysis membrane should be chosen according to the sequence of DNA used.
5. Generally, tris(2-carboxyethyl)phosphine is used to break any disulfide bonds in thiolated DNA. If there is no –SH in DNA, the process is not needed.
6. The ECL method is highly sensitive method, in which the ECL intensity is affected by various conditions, such as the sensitivity of the ECL instrument and the area of working electrode, the distance between the working electrode and the detector, etc. Careful operation should be done.

Acknowledgments

This work was supported by National Natural Science Foundation of China (Nos. 21375102, 21775097).

References

1. Watson JD, Crick F (1953) Molecular structure of nucleic acids; a structure for deoxyribose nucleic acid. Nature 171:737–738
2. Jones PA, Takai D (2001) The role of DNA methylation in mammalian epigenetics. Science 293:1068–1070
3. Münzel M, Globisch D, Brückl T, Wagner M, Welzmiller V, Michalakis S, Müller M, Biel M, Carell T (2010) Quantification of the sixth DNA base hydroxymethylcytosine in the brain. Angew Chem Int Ed 49:5375–5377
4. Kriaucionis S, Heintz N (2009) The nuclear DNA base 5-hydroxymethylcytosine is present in Purkinje neurons and the brain. Science 324:929–930
5. Song CX, Clark TA, Lu XY, Kislyuk A, Dai Q, Turner SW, He C, Korlach J (2012) Sensitive and specific single-molecule sequencing of 5-hydroxymethylcytosine. Nat Methods 9:75–77
6. Chen HY, Wei JR, Pan JX, Zhang W, Dang FQ, Zhang ZQ, Zhang J (2017) Spectroscopic quantification of 5-hydroxymethylcytosine in genomic DNA using boric acid-functionalized nano-microsphere fluorescent probes. Biosens Bioelectron 91:328–333
7. Chen SX, Dou YZ, Zhao ZH, Li FW, Su J, Fan CH, Song SP (2016) High-sensitivity and high-efficiency detection of DNA hydroxymethylation in genomic DNA by multiplexing electrochemical biosensing. Anal Chem 88:3476–3480
8. Ma SX, Sun HP, Li Y, Qi HL, Zheng JB (2016) Discrimination between 5-hydroxymethylcytosine and 5-methylcytosine in DNA via selective electrogenerated chemiluminescence (ECL) labeling. Anal Chem 88:9934–9940
9. Song CX, Szulwach KE, Fu Y, Dai Q, Yi CQ, Li XK, Chen CH, Zhang W, Jian X, Wang J, Looney T, Zhang B, Godley LA, Hicks LM, Lahn BT, Jin P, He C (2011) Selective chemical labeling reveals the genome-wide distribution of 5-hydroxymethylcytosine. Nat Biotechnol 29:68–72
10. Booth MJ, Branco MR, Ficz G, Oxley D, Krueger F, Reik W, Balasubramanian S (2012) Quantitative sequencing of 5-methylcytosine and 5-hydroxymethylcytosine at single-base resolution. Science 336:934–937
11. Hong TT, Wang TL, Guo P, Xing XW, Ding F, Chen YQ, Wu JJ, Ma JW, Wu F, Zhou X (2013) Fluorescent strategy based on cationic conjugated polymer fluorescence resonance energy transfer for the quantification of 5-(hydroxymethyl)cytosine in genomic DNA. Anal Chem 85:10797–10802
12. Qi HL, Zhang CX (2020) Electrogenerated chemiluminescence biosensing. Anal Chem 92:524–534
13. Miao WJ (2008) Electrogenerated chemiluminescence and its biorelated applications. Chem Rev 108:2506–2553
14. Meng Y, Bai WQ, Zhang YL, Sun HP, Li Y (2020) Electrogenerated chemiluminescence biosensing method based on 5-hydroxymethylcytosine antibody and PDDA-CNTs nanocomposites for the determination of 5-hydroxymethylcytosine double-stranded DNA. Talanta 210:120597
15. Wang H, Liu MZ, Bai WQ, Sun HP, Li Y, Deng HQ (2019) A convenient electrogenerated chemiluminescence biosensing method for selective detection of 5-hydroxymethylcytosine in genomic DNA. Sensors Actuators B Chem 284:236–242
16. Sun HP, Zhang JW, Bai WQ, Li Y (2020) Electrogenerated chemiluminescence biosensing method for DNA hydroxymethylation detection via glycosylation and a new multifunctional ECL signal compound. Sensors Actuators B Chem 322:128582
17. Zhang YL, Li Y, Wei YY, Sun HP, Wang H (2017) A sensitive signal-off electrogenerated chemiluminescence biosensing method for the discrimination of DNA hydroxymethylation based on glycosylation modification and signal quenching from ferroceneboronic acid. Talanta 170:546–551
18. Sun HP, Wang H, Bai WQ, Bao L, Lin JX, Li Y (2019) Sensitive electrogenerated chemiluminescence biosensing method for the determination of DNA hydroxymethylation based on $Ru(bpy)_3^{2+}$–doped silica nanoparticles labeling and MoS_2-poly(acrylic acid) nanosheets modified electrode. Talanta 191:350–356

19. Qi HL, Zhang Y, Peng YG, Zhang CX (2008) Homogenous electrogenerated chemiluminescence immunoassay for human immunoglobulin G using N-(aminobutyl)-N-ethylisoluminol as luminescence label at gold nanoparticles modified paraffin-impregnated graphite electrode. Talanta 75:684–690

Chapter 8

Quantification of Site-Specific 5-Formylcytosine by Integrating Peptide Nucleic Acid-Clamped Ligation with Loop-Mediated Isothermal Amplification

Zhenhao Zhang, Weimin Tian, Wei Ren, Zhengping Li, and Chenghui Liu

Abstract

The precise sensing of 5-formylcytosine (5fC) is of great significance because 5fC not only determines the nucleosome organization that regulates gene expression but also acts as a potential epigenetic marker for cancers. Nevertheless, owing to the low abundance of 5fC in genomic DNA as well as the strong interference signals introduced by its analogous including cytosine (C), 5-methylcytosine (5mC), 5-hydroxymethylcytosine (5hmC), and 5-carboxylcytosine (5caC), the high-sensitive quantification of site-specific 5fC remains challenging. To solve this issue, we have developed a new strategy that integrates peptide nucleic acid (PNA)-clamped ligation with loop-mediated isothermal amplification (LAMP), which is feasible and reliable for the determination of site-specific 5fC in genomic DNA. In this chapter, the detailed protocols for discriminating 5fC from its analogous, through 5fC-specific signal amplification are described. These protocols may also be applicable to the high-sensitive detection of other nucleic acid biomarkers.

Keywords 5-Formylcytosine, Peptide nucleic acid, Ligation, Loop-mediated isothermal amplification, Background-free quantification

1 Introduction

The 5-formylcytosine (5fC) has long been regarded as a transient compound in the process of DNA demethylation [1–4]. However, it is newly discovered that 5fC is a stable nucleotide that possesses distinct biological functions in genomic DNA [5–8]. For example, recently it has been revealed that 5fC is not only a key factor in nucleosome organization to regulate gene expression but also a possible epigenetic biomarker of cancers [9–14]. In this regard, the determination and quantification of site-specific 5fC are critical to further uncovering its underlying physiological and pathological functions. Nevertheless, there are two main roadblocks on the way to the accurate detection of 5fC. On the one hand, the low abundance of 5fC in genomic DNA requires ultrahigh sensitivity of the

Bi-Feng Yuan (ed.), *DNA Modification Detection Methods*, Springer Protocols Handbooks,
https://doi.org/10.1007/978-1-0716-1229-3_8,

detection approaches. On the other hand, 5fC behaves like its analogues including cytosine (C), 5-methylcytosine (5mC), 5-hydroxymethylcytosine (5hmC), and 5-carboxylcytosine (5caC) in base pairing, which will induce inevitable interference signals in 5fC quantification.

Up to now, great efforts have been devoted to developing simple and accurate detection methods to distinguish 5fC at single-base resolution [15–19]. For example, the reduced bisulfite sequencing and 5fC-assisted bisulfite sequencing allow to map 5fC in genomic DNA [20, 21]. But both the strategies are based on bisulfite treatment, which will lead to DNA degradation because of the harsh acidic and thermal conditions [22, 23]. Recently, a series of elegant bisulfite-free sequencing strategies have been developed by labeling 5fC with specific compounds [24, 25]. Although these strategies have enabled the mapping of 5fC under mild conditions, their routine clinical use remains hindered because they are costly and time-consuming.

To solve the two key challenges in specificity and sensitivity, a new strategy that innovatively integrates peptide nucleic acid (PNA)-clamped ligation with loop-mediated isothermal amplification (LAMP) has been proposed for the site-specific determination of 5fC [26]. In the initial step, malononitrile is employed to specifically convert 5fC to 5fC-M (M denotes malononitrile) in high efficiency, meanwhile, the other analogous of 5fC keep unchanged [25]. It should be noted that malononitrile will not influence the subsequent amplification process in 5fC quantification. After then, PNA will tightly clamp the 5fC analogous-containing DNA molecules by base pairing, whereas the 5fC-M-containing ones remain unclamped because 5fC-M behaves like thymine (T) in complementary hybridization [27, 28]. Subsequently, only the 5fC-containing DNA molecules are able to mediate the next-step ligation reaction to yield double stem-loop DNA structures as the initiative materials for LAMP, and the C/5mC/5hmC/5caC-containing DNA molecules generate none because they are clamped by PNA. In this way, the interference of 5fC's analogues is efficiently suppressed because both the chemical conversion by malononitrile and PNA clamping are of high efficiencies. At last, the 5fC-induced initiative materials will trigger the LAMP to amplify the fluorescent signal in an exponential manner. As a result, the quantification of ultralow concentration of 5fC in a background-free fashion is achieved by adopting this strategy. In this chapter, the detailed protocols for discriminating 5fC from its analogous based on the combination of PNA-clamped ligation and LAMP are described.

2 Materials

All the materials employed in the PNA-clamped ligation LAMP strategy are listed below.

2.1 Specific Chemical Conversion of 5fC to 5fC-M

2.1.1 Sequence Information

Oligonucleotide DNA (ODN-DNA) sequences are designed according to Lrpap1 gene in chromosome 5 of Mus musculus strain C57BL/6J. The underlined bases are the target sites in the sequences. The sequences shown in Table 1 are synthesized and purified by TaKaRa Biotechnology (Dalian, China). The as-received products are colorless DNA dry powder, which need to be dissolved to obtain 10μM solution by using nuclease-free water and stored at −80 °C.

2.1.2 Reagents and Solution

1. 500 mM NH_4Ac: the pH of the solution is adjusted to 7.0 with ammonium hydroxide.
2. 100 mM malononitrile (Sigma-Aldrich): dissolved in 500 mM NH_4Ac (*see* **Note 1**).

2.1.3 Equipment

Shaking drybath (Crystal Technology & Industries, Inc).

2.2 Distinguish 5fC from Its Analogues by PNA-Clamped Ligation Reaction

2.2.1 Sequence Information

Probe A, Probe B, front inner primer (FIP), and backward inner primer (BIP) (Table 2) are synthesized and purified by TaKaRa Biotechnology (Dalian, China). The as-received products are also colorless DNA dry powder, which need to be dissolved to obtain 10μM solution by using nuclease-free water, and stored at −20 °C. The PNA is synthesized and purified by Panagene (Daejeon, Korea, *see* **Note 2**). The product we obtained is also colorless dry powder, which is dissolved into 100 μM solution with nuclease-free water and stored at −80 °C. Italics sequences of probes are complementary to the target sequences (ODN-DNA) and can be replaced by other target-specific sequences for 5fC site determination. Probe A is modified with a phosphate group at its 5′-end. The nucleotide on the 3′-end (also the ligation site) of Probe B is adenine (A) to pair with 5fC-M. PNA sequence is complementary with the middle part of ODN-C, in which the underlined guanine (G) is the complementary site of C. FIP and BIP are universal primers for the LAMP reaction.

2.2.2 Reagents and Solution

1. 10× Amp ligase buffer (Epicentre Biotechnologies): 200 mM Tris-HCl, 250 mM KCl, 100 mM $MgCl_2$, 5 mM nicotinamide adenine dinucleotide (NAD), and 0.1% Triton X-100, pH 8.3.
2. 100 μM PNA (Panagene Inc).
3. 10 μM Probe A and Probe B.
4. Amp ligase (Epicentre Biotechnologies): 5 U/μL (*see* **Note 3**).
5. Nuclease-free water (TaKaRa Biotechnology).

Table 1
The sequence of targets

Name	Sequence (5′–3′)
ODN-T	CCAGGTCCCACAGATCTATCACCTGGGGCTCTTCAAACTCTGCAGG
ODN-C	CCAGGTCCCACAGATCTATCACCCGGGGCTCTTCAAACTCTGCAGG
ODN-5mC	CCAGGTCCCACAGATCTATCACC5mCGGGGCTCTTCAAACTCTGCAGG
ODN-5hmC	CCAGGTCCCACAGATCTATCACC5hmCGGGGCTCTTCAAACTCTGCAGG
ODN-5fC	CCAGGTCCCACAGATCTATCACC5fCGGGGCTCTTCAAACTCTGCAGG
ODN-5caC	CCAGGTCCCACAGATCTATCACC5caCGGGGCTCTTCAAACTCTGCAGG
Complementary sequence of ODN-C	CCTGCAGAGTTTGAAGAGCCCCGGGTGATAGATCTGTGGGACCTGG
Fragment sequence of genome DNA	TCCTTACCCTGAATGACTCCAGTTCCTTCTCAGTGAAGTTGGCAGACTGAGCCAGG TCCCACAGATCTATCACCCGGGGCTCTTCAAACTCTGCAGGAGAGCAAGGGCTGTCTA TAGGTGGCAAGTCAGGACACACCTCGGTGCTGGGGGCCAAGTGGCTTGTCA TCCCAAGGGAATAGTGTGGGTTG Complementary sequence CAACCCACACTATTCCCTTGGGATGACAAGCCACTTGGCCCCCAGCACCGAGGTGTGTCC TGACTTGCCACCTATAGACAGCCCTTGCTCTCCTGCAGAGTTTGAAGAGCCCCGGGTGA TAGATCTGTGGGACCTGGCTCAGTCTGCCAACTTCACTGAGAAGGAACTGGAGTCA TTCAGGGTAAGGA

Table 2
The sequence of probes and primers for LAMP

Name	Sequence (5′–3′)
Probe A	Phosphate-*GGTGATAGATCTGTGGAC*TTTTGTCTGGCAGTGTGTTT CTATATCGGTCCTCTCCTCCTTTCACACTGCCAGAC
Probe B	GCCTACTACTTCGTTTCTTGCCTTTGCTCTCTACTGACTTTTCGAAGTAGTAGGC TTTT*GAGTTTGAAGAGCCCCA*
PNA	CCGGGTGATAG
FIP	GTCTGGCAGTGTGAAAGGAGGAGAGAGGACCGATATAG
BIP	GCCTACTACTTCGTTTCTTGCCTTTGCTCTCTACTGAC

2.2.3 Equipment and Consumables

1. 2720 thermal cycler (Applied Biosystems).
2. PCR tube (Axygen): 0.2 mL.
3. Pipet tips (Axygen): 0.2 mL; pipet tips (Thermo QSP): 10 μL.
4. Pipette (Dragon): 0.5–10 μL, 20–200 μL.

2.3 Signal Amplification by LAMP

2.3.1 Reagents and Solution

1. 10× ThermoPol reaction buffer (New England Biolabs): 200 mM Tris-HCl, 100 mM KCl, 20 mM $MgSO_4$, 100 mM $(NH_4)_2SO_4$, and 0.1% Triton X-100, pH 8.8.
2. 5 M betaine (Sigma-Aldrich): dissolved in nuclease-free water (*see* **Note 4**).
3. 10 μM FIP, 10 μM BIP (TaKaRa Biotechnology, detailed sequence can be found in Subheading 8.2.2.1).
4. Bst DNA polymerase (New England Biolabs): 8 U/μL.
5. SYBR Green I (Xiamen Bio-Vision Biotechnology): 20× stock solution in DMSO.
6. dNTP mixture (TaKaRa Biotechnology): 2.5 mM dATP, dTTP, dCTP, and dGTP, respectively.

2.3.2 Equipment and Consumables

1. StepOne Real-Time PCR System (Applied Biosystems).
2. 8-tube strips (Applied Biosystems): 0.1 mL.
3. Optical 8-cap strips (Applied Biosystems): 0.1 mL.

2.4 Polyacrylamide Gel Electrophoresis (PAGE) Analysis

2.4.1 Reagents and Equipment

1. 20% ammonium persulfate (APS, Sigma-Aldrich): 0.2 g of APS dissolved in 800 μL nuclease-free water.
2. Tris-borate-EDTA (TBE) electrophoresis buffer: 9 mM Tris-HCl, 9 mM boric acid, 0.2 mM EDTA, pH 7.9.
3. Tetramethylethylenediamine (TEMED, Sigma-Aldrich).
4. Acryl/Bis 40% solution (29:1, Shanghai Sangon Biological Engineering Technology & Services Co., Ltd).

5. SYBR™ Gold Nucleic Acid Gel Stain (Life Science).
6. Loading buffer (6×) (TaKaRa Biotechnology).
7. 20 bp DNA ladder (Dye Plus) (TaKaRa Biotechnology).
8. Electrophoresis apparatus (Beijing Liuyi Biotechnology Co., Ltd.)
9. Gel Doc EZ imager (Bio-Rad).

3 Methods

An overview of all the procedures involved in the PNA-clamped ligation LAMP is indicated in Fig. 1. The procedures involve three parts, including malononitrile conversion reaction, PNA-clamped ligation reaction, and LAMP reaction (*see* **Note 5**). The reagents required and the time consumed in these reactions are listed below the corresponding steps. The whole process takes about 21 h (including the overnight chemical conversion, 8 samples handled in parallel).

3.1 Malononitrile-Based 5fC-Specific Conversion Reaction

To distinguish 5fC from its analogue X (X represents C, 5mC, 5hmC, or 5caC) which has the similar structure and base-pairing behavior to 5fC, in this strategy the malononitrile-based chemical conversion that specifically converts 5fC to 5fC-M is carried out. Malononitrile is a stable and commercially available reagent that has excellent water solubility. It can specifically convert 5fC to 5fC-M, while leaving other 5fC analogues unaffected (Fig. 2) [25, 26]. 5fC-M behaves like T in base pairing, hence, the difference between 5fC and its analogues is converted into the difference between T and C, which can be directly distinguished in nucleic acid base pairing. The detailed steps of the conversion reaction are as follows:

1. Prepare conversion reaction solution: 100 mM malononitrile in 500 mM NH_4Ac solution (pH 7.0).
2. Prepare detection samples: the determination of site-specific 5fC is conducted on three kinds of samples, including ODN-DNA, double-strand DNA sequence (dsDNA), and genomic DNA samples.
   - ODN-DNA sample: dilute 10μM ODN-C, ODN-5mC, ODN-5hmC, ODN-5fC, and ODN-5caC stock solution to 1 μM by using nuclease-free water (*see* **Note 6**). Add 1μL of the ODN-DNA solution (1 μM) to the conversion reaction solution, respectively.
   - dsDNA sample: add 1μL ODN-DNA (1 μM) and 1μL complementary sequence of ODN-C (1 μM) to the conversion reaction solution and then incubate the mixture at 90 °C for 3 min, then gradually cool down to room temperature to form dsDNA.

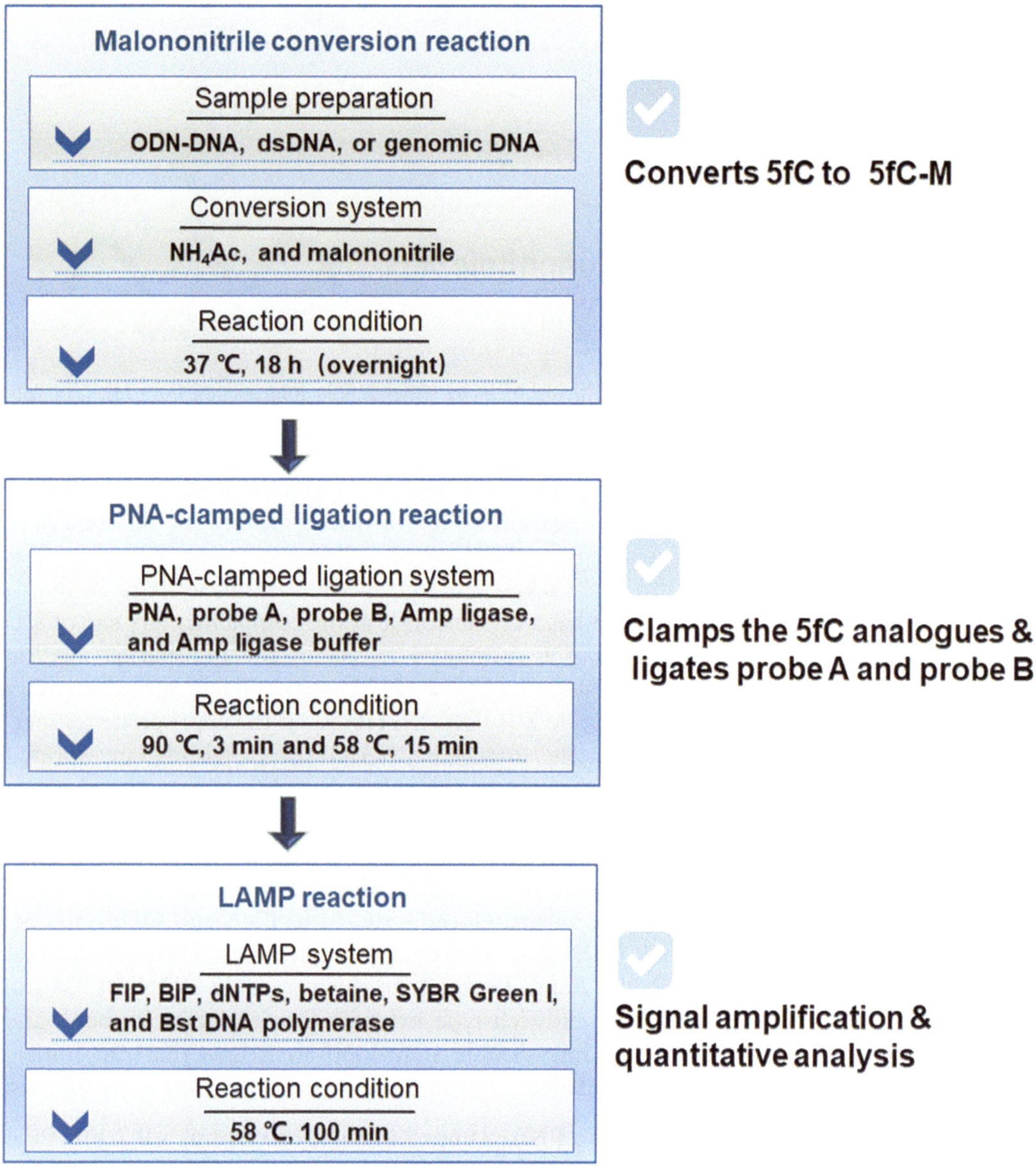

Fig. 1 Workflow of the site-specific 5fC detection strategy

- Genomic DNA samples: extract the genomic DNA from specific tissues of interest (taking mouse brain tissue as an example in this protocol). Add an appropriate amount of genomic DNA to the conversion reaction solution.

3. Add nuclease-free water to the conversion reaction mixture for a 10μL total volume.
4. Mix the 10μL reaction system thoroughly with a pipette.
5. Incubate the mixture at 37 °C for 18 h (overnight) on a Shaking drybath (850 rpm).

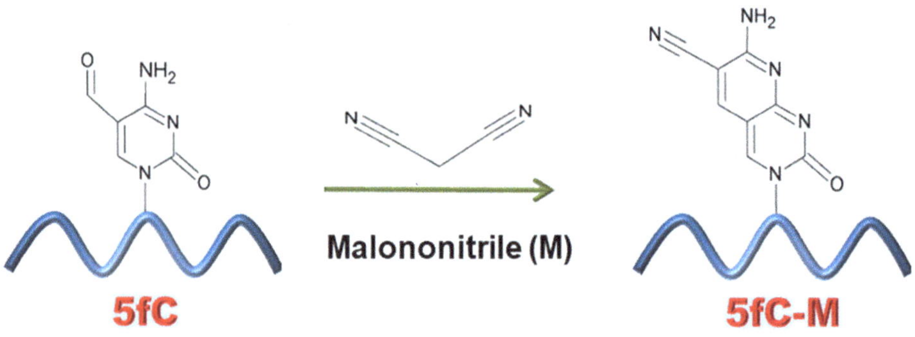

Fig. 2 The specific chemical conversion of 5fC to 5fC-M by malononitrile (reproduced from ref. [26] with permission from American Chemical Society)

3.2 PNA-Clamped Ligation Reaction

Nonspecific ligation is inevitable in the previous ligation-based methods (e.g., ligation PCR) for distinguishing the differences of single bases by ligase [29], which will lead to nonspecific amplification to produce background interference. In this strategy, an 11-nt-long PNA that has a G nucleotide at the complementary site of 5fC or its analogous is employed to clamp the nonspecific ligation reaction. As shown in Fig. 3, PNA is completely complementary to oligonucleotide X (ODN-X). Hence, at the appropriate temperature, completely matched PNA:ODN-X hybrids are formed because PNA can clamp the ODN-X tightly. Thus, the clamped ODN-X can no longer induce the nonspecific ligation of Probe A and B, forbidding the producing of the initiative materials for LAMP. In contrast, ODN-5fC-M remains unclamped owing to the single-base mismatch of G:5fC-M in PNA and ODN-5fC-M. Therefore, ODN-5fC-M is capable of efficiently bridging the ligation of Probe A and Probe B. In this way, only ODN-5fC-M can template the sandwich-type hybridization to ligate Probe A and Probe B, yielding double stem-loop structures that can initiate LAMP. The detailed steps of the conversion reaction are as follows:

1. Prepare the PNA-clamped ligation mixture in a 0.2 mL PCR tube. The reaction mixture contains 300 nM PNA, 2 nM Probe A, 2 nM Probe B, 1× Amp ligase buffer, and 0.02 U Amp ligase (*see* **Note** 7).
2. Add 1μL certain concentration of targets to the ligation mixture.
3. Add nuclease-free water to the ligation mixture to make a 10μL total volume and gently mix with a pipette.
4. Incubate the mixture at 90 °C for 3 min to denature the dsDNA or genomic DNA sequence, and then incubate at 58 °C for 15 min to ligate Probe A and Probe B at 2720 thermal cycler (*see* **Note 8**).

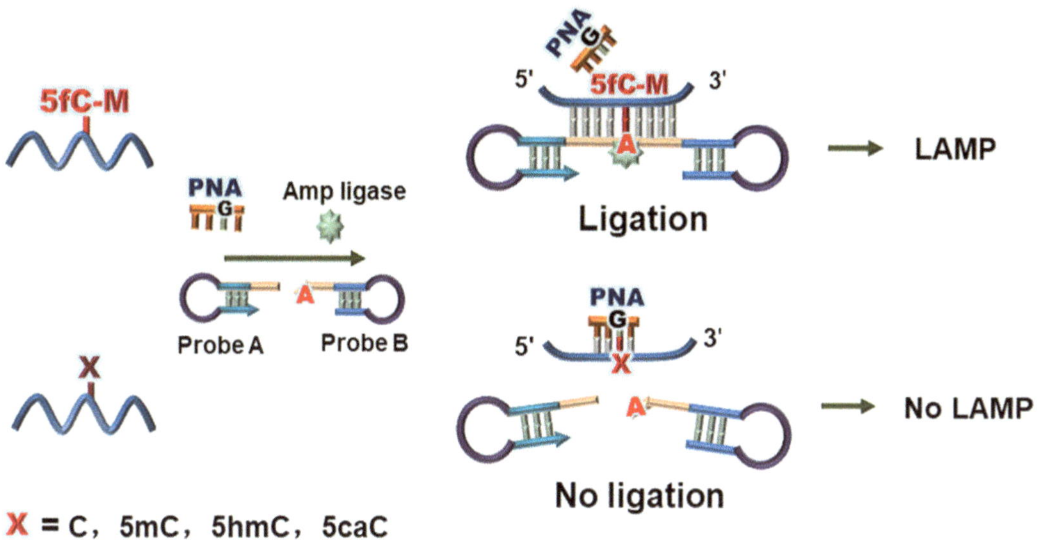

Fig. 3 Schematic representation of the PNA-clamped ligation reaction (reproduced from ref. [26] with permission from American Chemical Society)

3.3 LAMP Reaction

LAMP is an isothermal amplification approach that can rapidly amplify the ligation product bridged by ODN-5fC with high specificity and efficiency [30]. As shown in Fig. 4, the double stem-loop structure produced by the ligation reaction is named Structure I. Structure I is the essential initiative material for LAMP. FIP and BIP are the primers for the amplification reaction. On the stem-loop structure, the fragments named Fs and BsC can hybridize with FIP and BIP, respectively ($s = 1$ or 2). Meanwhile, FsC and BsC are the complementary sequences of Fs and Bs. With the help of Bst DNA polymerase and dNTPs, Structure I self-extends from its 3′-end, forming Structure II. Afterward, FIP hybridizes with the F1 loop on Structure II and then extends along its 3′ terminal to open the double-strand stem of Structure II. Simultaneously, a new loop structure forms on the released 3′-tail of Structure II, generating Structure III. With the B2C of Structure III extends along the 3′ direction, the extended product of FIP will be displaced, releasing Structure I^*. The original stem-loop structure is named as Structure IV. As the self-extension of Structure I^* (along the 3′ direction) as well as the hybridization of BIP and B1C loop going on, a new cycle of internal polymerization is initiated, continuously producing Structure I and Structure VII, respectively. The polymerization initiated by FIP and BIP not merely works on Structure I and Structure I^*, but also works on the stem-loop structure of amplification products, resulting in the exponentially amplified dsDNA as the self-powered polymerization repeats continuously.

Since LAMP is an efficient exponential amplification mechanism, real-time fluorescence analysis can be used to read the

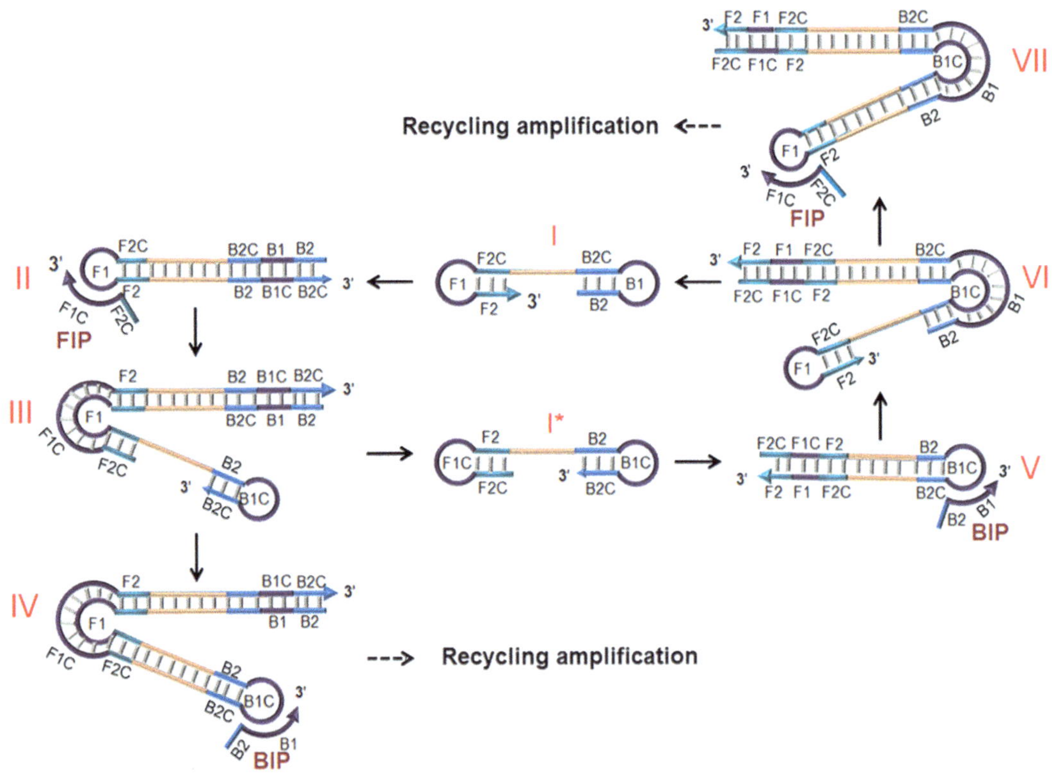

Fig. 4 Schematic representation of LAMP reaction (reproduced from ref. [26] with permission from American Chemical Society)

amplified signals with the assistance of SYBR Green I which is a dsDNA-specific fluorescent dye. The point of inflection (POI) value is introduced for the target quantitation which represents the time corresponding to the maximum slope in each real-time fluorescence curve. In this process, the higher the ODN-5fC level is, the more initiative material of LAMP would be produced, thus making the real-time fluorescence curve slope appear earlier and the POI value smaller. The detailed steps of LAMP are as follows:

1. Prepare LAMP reaction mixture that contains 1× ThermoPol reaction buffer, 0.4μM FIP, 0.4μM BIP, 200μM dNTPs, 1 M betaine, 0.4× SYBR Green I, and 4 U Bst DNA polymerase.
2. Add 2μL of ligation products to the LAMP reaction mixture (*see* **Note 9**).
3. Add nuclease-free water to the LAMP reaction mixture to make a 10μL total volume and mix thoroughly.
4. Transfer the 10μL mixture to 8-tube strips and then cover with optical 8-cap strips.
5. Put 8-tube strips in the StepOne Real-Time PCR System to perform the LAMP reaction (*see* **Note 10**).

6. Set the LAMP reaction system as 58 °C for 100 min, and the real-time fluorescence intensity is monitored at 1-min intervals.

3.4 PAGE Characterization of the Amplification Products of ODN-5fC and ODN-X

The specificity of discriminating 5fC from its analogues is imperative for the PNA-clamped ligation LAMP assay. In this method, only ODN-5fC should be able to induce the specific ligation and the subsequent LAMP to produce large amounts of long double-strands. However, ODN-X cannot produce detectable amplification signals. To guarantee the specificity for the detection of 5fC-containing DNA, the amplification products of ODN-5fC and ODN-X are characterized by PAGE. The detailed steps of PAGE analysis are as follows:

1. Prepare 16% PAGE mixture: 1.9 mL H_2O, 4 mL Acryl/Bis 40% solution (29:1), 4 mL TBE buffer, 50μL 20% APS, 7 μL TEMED. Pour the mixture into the glass channel, leave it at room temperature for 40 min.
2. Add 4 μL LAMP products and 1 μL loading buffer (6×) to 0.2 mL PCR tube, mix thoroughly and then load the sample onto PAGE gel.
3. Add 20 bp DNA ladder in one well of PAGE gel.
4. Maintain 110 V constant voltage for 80 min in electrophoresis apparatus at room temperature.
5. Dilute SYBR™ Gold Nucleic Acid Gel Stain 10,000-fold with TBE buffer to make a 1× staining solution.
6. Stain the PAGE gel with 1× staining solution for 5 min.
7. Take image of PAGE gel and analyze the sample with a Gel Doc EZ imager.

The PAGE image is shown in Fig. 5. "+M" indicates that the samples are treated with malononitrile, while "+PNA" indicates that PNA is present in the ligation mixture. Lane 1 contains DNA markers (20 bp DNA ladder). Bands below 40 bp DNA ladder represent unreacted FIP and BIP primers. The ladder-like bands in lane 5 represent the characteristic pattern of LAMP products for ODN-5fC. In contrast, ODN-X has shown none except the primers, indicating that there are no amplification products. These phenomena prove the assay can remove the background interference and is of high specificity to distinguish 5fC-containing DNA from its analogous.

3.5 Quantitative Analysis of 5fC

The level of dsDNA produced by the PNA-clamped ligation LAMP can be used to quantitatively analyze 5fC in ODN-5fC, dsDNA-5fC, or genomic DNA.

1. Dilute the converted products of 1μM ODN-5fC or 1μM dsDNA-5fC to 5 pM, 2.5 pM, 500 fM, 250 fM, 50 fM, 10

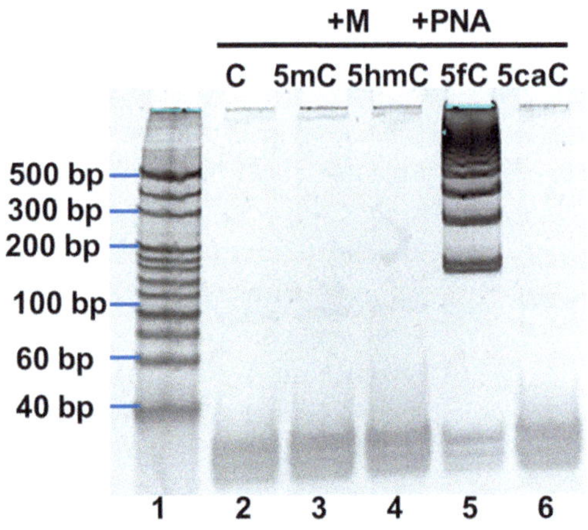

Fig. 5 Lane 1: DNA markers (20 bp DNA ladder); Lane 2, 3, 4, 5, 6 are LAMP products of ODN-C, ODN-5mC, ODN-5hmC, ODN-5fC, and ODN-5caC, respectively (reproduced from ref. [26] with permission from American Chemical Society)

fM and then perform the PNA-clamped ligation LAMP (the steps of Subheadings 8.3.1, 8.3.2, and 8.3.3).

2. The data of each sample exported from StepOne Real-Time PCR system is analyzed. The real-time fluorescence curves are obtained by taking time as the abscissa and fluorescence intensity as the ordinate.
3. By analyzing the maximum slope of the real-time fluorescence curve, the corresponding time which is POI value can be obtained. The mean value and standard deviation of the POI values are obtained from three measurements of the sample.
4. Set the mean POI value as the ordinate, and the negative logarithm value of the concentration of 5fC ($-\lg C$) in the corresponding sample as the abscissa, the standard curve, and the corresponding correlation equation can be obtained.

3.5.1 Quantitative Analysis of 5fC in ODN-5fC and dsDNA-5fC

According to the design of the PNA-clamped ligation LAMP, the higher the concentration of the converted ODN-5fC added to the PNA-clamped ligation mixture, the earlier the real-time fluorescence curve appears, and the smaller the POI value is. There would be a linear relationship between the POI values and $\lg C$ in the range of 10^{-16} to 10^{-13} M. In addition, the analytical performance for detecting dsDNA-5fC is in good accordance with that for ODN-5fC [26]. This strategy allows for the determination of 5fC in dsDNA without the interference from the complementary sequences.

3.5.2 Quantitative Analysis 5fC in Mouse Brain Tissue Genomic DNA

To assess the practicability of the PNA-clamped ligation LAMP, in this protocol it is applied to the determination of site-specific 5fC in Lrpap1 gene of mouse brain genomic DNA.

1. Extract genome DNA from the brain tissues of two 3-week healthy mice.
2. The PNA-clamped ligation LAMP is performed on 120 ng genome DNA samples extracted from two healthy mice, respectively.
3. According to the corresponding correlation equation between the dosages of standard dsDNA-5fC and the POI values, the concentrations of site-specific 5fC-containing DNA in the two genomic DNA samples can be determined.

4 Note

These notes highlight a few critical points for each step presented above.

1. Malononitrile is a toxic compound by all toxic routes. It should be stored in a cool and ventilated storeroom and kept away from fire and heat. The package must be sealed and not in contact with air. PPE (gloves, face mask, and eye mask) should be worn when taking malononitrile.
2. Procedures and the note for PNA sequence design can be found on the web page: http://www.tahepna.com/tool2.asp?flag=6.
3. Amp ligase is an NAD-dependent DNA ligase. Amp ligase and Amp ligase buffer need to be placed at −80 °C for long-term storage.
4. Betaine solution needs to be stored at −20 °C.
5. In order to prevent the operator from contaminating the enzyme and reaction mixture, a mask and gloves should be worn throughout the operation. Moreover, the steps of Subheadings 8.3.2 and 8.3.3 contain enzyme reactions, hence, these processes should be performed on ice.
6. The target sequences are stored in −80 °C and need to be thawed at 4 °C before use.
7. When preparing a reaction mixture with enzymes, the enzymes should be added in the end of the operation.
8. In the PNA-clamped ligation reaction, the optimal ligation temperature may vary with the changes in the PNA sequence and target-specific sequence of probes. Therefore, the ligation temperature needs to be optimized when detecting 5fC at different sites.

9. After the ligation reaction, the ligation products should be first placed on ice for 5 min before introducing to the subsequent LAMP reaction component.
10. The LAMP mixture must be quickly transferred to 8-tube strips and then immediately covered with the 8-cap strips. Quickly put the samples into the StepOne Real-Time PCR System and perform the isothermal amplification program.

References

1. He YF, Li BZ, Li Z, Liu P, Wang Y et al (2011) Tet-mediated formation of 5-carboxylcytosine and its excision by TDG in mammalian DNA. Science 333:1303–1307
2. Ito S, Shen L, Dai Q, Wu SC, Collins LB et al (2011) Tet proteins can convert 5-methylcytosine to 5-formylcytosine and 5-carboxylcytosine. Science 333:1300–1303
3. Zhang L, Chen W, Iyer LM, Hu J, Wang G et al (2014) A TET homologue protein from Coprinopsis cinerea (CcTET) that biochemically converts 5-methylcytosine to 5-hydroxymethylcytosine, 5-formylcytosine, and 5-carboxylcytosine. J Am Chem Soc 136:4801–4804
4. Wagner M, Steinbacher J, Kraus TF, Michalakis S, Hackner B et al (2015) Age-dependent levels of 5-methyl-, 5-hydroxymethyl-, and 5-formylcytosine in human and mouse brain tissues. Angew Chem Int Ed 54:12511–12514
5. Bachman M, Uribe-Lewis S, Yang X, Burgess HE, Iurlaro M et al (2015) 5-formylcytosine can be a stable DNA modification in mammals. Nat Chem Biol 11:555–557
6. Iurlaro M, McInroy GR, Burgess HE, Dean W, Raiber EA et al (2016) In vivo genome-wide profiling reveals a tissue-specific role for 5-formylcytosine. Genome Biol 17:R114
7. Iurlaro M, Ficz G, Oxley D, Raiber EA, Bachman M et al (2013) A screen for hydroxymethylcytosine and formylcytosine binding proteins suggests functions in transcription and chromatin regulation. Genome Biol 14:R119
8. Raiber EA, Murat P, Chirgadze DY, Beraldi D, Luisi BF et al (2015) 5-formylcytosine alters the structure of the DNA double helix. Nat Struct Mol Biol 22:44–49
9. Raiber EA, Portella G, Martinez Cuesta S, Hardisty R, Murat P et al (2018) 5-formylcytosine organizes nucleosomes and forms Schiff base interactions with histones in mouse embryonic stem cells. Nat Chem 10:1258–1266
10. Wang Y, Zhang X, Zou G, Peng S, Liu C et al (2019) Detection and application of 5-formylcytosine and 5-formyluracil in DNA. Acc Chem Res 52:1016–10244
11. Tang Y, Zheng SJ, Qi CB, Feng YQ, Yuan BF (2015) Sensitive and simultaneous determination of 5-methylcytosine and its oxidation products in genomic DNA by chemical derivatization coupled with liquid chromatography-tandem mass spectrometry analysis. Anal Chem 87:3445–3452
12. Liu J, Jiang J, Mo J, Liu D, Cao D et al (2019) Global DNA 5-hydroxymethylcytosine and 5-formylcytosine contents are decreased in the early stage of hepatocellular carcinoma. Hepatology 69:196–208
13. Storebjerg TM, Strand SH, Høyer S, Lynnerup AS, Borre M et al (2018) Dysregulation and prognostic potential of 5-methylcytosine (5mC), 5-hydroxymethylcytosine (5hmC), 5-formylcytosine (5fC), and 5-carboxylcytosine (5caC) levels in prostate cancer. Clin Epigenetics 10:105
14. Jiang HP, Liu T, Guo N, Yu L, Yuan BF et al (2017) Determination of formylated DNA and RNA by chemical labeling combined with mass spectrometry analysis. Anal Chim Acta 981:1–10
15. Gieß M, Muñoz-López A, Buchmuller B, Kubik G, Summerer D (2019) Programmable protein-DNA cross-linking for the direct capture and quantification of 5-formylcytosine. J Am Chem Soc 141:9453–9457
16. Liu C, Wang Y, Yang W, Wu F, Zeng W et al (2017) Fluorogenic labeling and single-base resolution analysis of 5-formylcytosine in DNA. Chem Sci 8:7443–7447
17. Wang Y, Liu C, Zhang X, Yang W, Wu F et al (2018) Gene specific-loci quantitative and single-base resolution analysis of 5-formylcytosine by compound-mediated polymerase chain reaction. Chem Sci 9:3723–3728
18. Wang SR, Song YY, Wei L, Liu CX, Fu BS et al (2017) Cucurbit[7]uril-driven host-guest chemistry for reversible intervention of 5-formylcytosine-targeted biochemical reactions. J Am Chem Soc 139:16903–16912

19. Wang Y, Liu C, Yang W, Zou G, Zhang X et al (2018) Naphthalimide derivatives as multifunctional molecules for detecting 5-formylpyrimidine by both PAGE analysis and dot-blot assays. Chem Commun 54:1497–1500
20. Booth MJ, Marsico G, Bachman M, Beraldi D, Balasubramanian S (2014) Quantitative sequencing of 5-formylcytosine in DNA at single-base resolution. Nat Chem 6:435–440
21. Song CX, Szulwach KE, Dai Q, Fu Y, Mao SQ et al (2013) Genome-wide profiling of 5-formylcytosine reveals its roles in epigenetic priming. Cell 153:678–691
22. Bai D, Yi C (2019) Advances in the profiling of single-cell DNA modifications. Small Methods 3:1900137
23. Tanaka K, Okamoto A (2007) Degradation of DNA by bisulfite treatment. Bloorg Med Chem Lett 17:1912–1915
24. Xia B, Han D, Lu X, Sun Z, Zhou A et al (2015) Bisulfite-free, base-resolution analysis of 5-formylcytosine at the genome scale. Nat Methods 12:1047–1050
25. Zhu C, Gao Y, Guo H, Xia B, Song J et al (2017) Single-cell 5-formylcytosine landscapes of mammalian early embryos and ESCs at single-base resolution. Cell Stem Cell 20:720–731
26. Zhang Z, Yang D, Tian W, Qi Y, Ren W et al (2020) Facile clamp-assisted ligation strategy for direct discrimination and background-free quantification of site-specific 5-formylcytosine. Anal Chem 92:3477–3482
27. Chiou CC, Luo JD, Chen TL (2006) Single-tube reaction using peptide nucleic acid as both PCR clamp and sensor probe for the detection of rare mutations. Nat Protoc 1:2604–2612
28. Luo JD, Chan EC, Shih CL, Chen TL, Liang Y et al (2006) Detection of rare mutant K-ras DNA in a single-tube reaction using peptide nucleic acid as both PCR clamp and sensor probe. Nucleic Acids Res 34:e12
29. Zhang Z, Shan X, Zhang P, Liu W, Yan J et al (2019) Highly sensitive quantification of site-specific 5-hydroxymethylcytosine at single-base resolution by HpaII-mediated ligation PCR. Org Biomol Chem 17:9849–9853
30. Notomi T, Okayama H, Masubuchi H, Yonekawa T, Watanabe K et al (2000) Loop-mediated isothermal amplification of DNA. Nucleic Acids Res 28:e63

Chapter 9

Global DNA Methylation Analysis Using Methylcytosine Dioxygenase

Natsumi Taka and Wataru Yoshida

Abstract

Because the alteration of global DNA methylation levels is detected in various cancer cells, the global DNA methylation level is expected as a biomarker for cancer diagnostics. Bisulfite-based assays have been widely used as a traditional method for the quantification of global DNA methylation levels. High-performance liquid chromatography (HPLC)-based assay and enzyme-linked immunosorbent assay (ELISA) have also been used; however, these methods require time-consuming DNA treatments, such as bisulfite treatment, digestion, or immobilization of genomic DNA before analysis. To analyze global DNA methylation levels in simple steps, an enzymatic assay using Ten-eleven translocation (TET), which oxidizes 5-methylcytosine (5mC) on genomic DNA, was developed. Succinate is produced during 5mC oxidation; thus, the amount of succinate produced by TET-mediated oxidation is correlated with the global DNA methylation level of genomic DNA. In this chapter, we describe the details of the expression of TET in *Escherichia coli* and the measurement of the amount of succinate produced from the TET-mediated reaction. The TET-based assay detects global DNA methylation levels in at least 100 ng genomic DNA.

Keywords Epigenetics, Global DNA methylation, 5-Methylcytosine oxidation, Ten-eleven translocation (TET), Succinate

1 Introduction

DNA methylation is the most commonly recognized modification and plays an important role in the regulation of gene expression, genomic imprinting, and cell development [1–4]. In vertebrates, DNA methylation occurs at a cytosine of CpG dinucleotide [5-methylcytosine (5mC)]. The three major DNA methyltransferase family members (DNMT1, DNMT3A, and DNMT3B) have been identified in mammals. DNMT1 plays a crucial role in the maintenance of DNA methylation patterns, whereas DNMT3A and DNMT3B are responsible for the establishment of de novo DNA methylation [3–5]. Ubiquitin-like with PHD and RING finger domains 1 (UHRF1) is involved in the maintenance of DNA

Bi-Feng Yuan (ed.), *DNA Modification Detection Methods*, Springer Protocols Handbooks,
https://doi.org/10.1007/978-1-0716-1229-3_9,

methylation. UHRF1 binds to hemi-methylated CpG and induces the recruitment of DNMT1 to the hemi-methylated sites [6–8].

Ten-eleven translocation (TET), which catalyzes the oxidation of 5mC to 5-hydroxymethylcytosine (5hmC), 5-formylmethylcytosine (5fC), and 5-carboxymethylcytosine (5caC), was discovered in mammals (Fig. 1) [9, 10]. The TET family (TET1, TET2, and TET3) contains a core catalytic domain: the cysteine-rich domain and the double-stranded β-helix (DSBH) domain [11–14]. TET1 and TET3 retain a CXXC domain that recognizes unmethylated CpG [15]. Meanwhile, TET2 has lost the CXXC domain through evolution and encodes IDAX (known as CXXC4) [11, 16]. TET is a member of the Fe(II)/α-ketoglutarate (α-KG)-dependent oxygenase family; therefore, TET-mediated 5mC oxidation is accompanied by converting α-KG to succinate and releasing a molecule of CO_2 [17]. It has been known that ascorbic acid is used as a reducing agent to maintain the active Fe(II) status during 5mC oxidation [18]. Therefore, the presence of ascorbic acid enhances TET-mediated reactions both in vivo and in vitro [19–21].

Aberrant DNA methylation, hypermethylation of promoter regions of cancer-associated genes and global hypomethylation of genomic DNA, have been widely observed in various cancers [1, 22–24]. Global DNA hypomethylation induces genome instability and increases mutation [25, 26]. The alteration of DNA methylation has also been observed in several neuropsychiatric diseases [27]. In the human brain, the global DNA methylation level is increased in Alzheimer's disease (AD) [28–30]. Loss of global DNA methylation has been observed in genomic DNA from blood samples of patients with Parkinson's disease (PD) [31, 32]. Thus, the measurement of global DNA methylation levels can be utilized to assess cancer risk and neuropsychiatric diseases, such as AD and PD.

Bisulfite conversion analysis [33–36], high-performance liquid chromatography (HPLC)-based assay [37], enzyme-linked immunosorbent assay (ELISA) [38], and methyl acceptance assay [39–41] have been widely used as traditional methods for the detection of global DNA methylation levels. However, these methods require cumbersome steps, namely bisulfite treatment, digestion, or immobilization of genomic DNA before analysis. To determine global DNA methylation levels in a homogeneous system, we previously developed an assay based on bioluminescence resonance energy transfer (BRET) using methyl-CpG binding domain (MBD)-fused firefly luciferase (MBD-Fluc) [42, 43] and unmethylated CpG-binding domain (CXXC)-fused firefly luciferase (CXXC-Fluc) [44]. MBD-Fluc and CXXC-Fluc bind to methylated and unmethylated CpG sites of genomic DNA, respectively. BRET signals occurred between fluorescent DNA intercalating dye and MBD-Fluc or CXXC-Fluc depending on methylated and

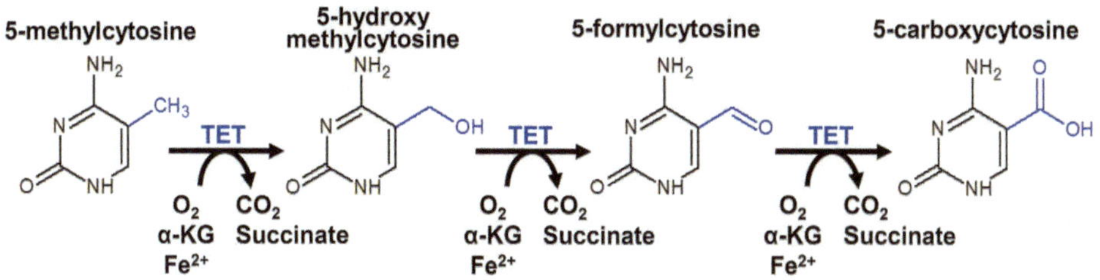

Fig. 1 The 5-methylcyrosine oxidation catalyzed by TET. The TET protein continuously oxidizes 5-methylcytosine to 5-hydroxymethylcytosine, 5-formylmethylcytosine, and 5-carboxymethylcytosine. α-KG is converted to succinate through oxidation

unmethylated CpG contents. Therefore, global DNA methylation levels can be quantified by these signals. Similarly, the amount of hemi-methylated CpG can be detected using SET and RING-associated (SRA) domain-fused firefly luciferase (SRA-Fluc), which binds to hemi-methylated CpG sites [45]. Using MBD-Fluc and CXXC-fused *Oplophorus* luciferase (Oluc), which is different from the peak emission of firefly luciferase, methylated and unmethylated CpG contents were detected using a multicolor assay in a single tube [46].

In this chapter, we describe a global DNA methylation sensing system using TET protein [47]. The amount of succinate produced by TET-mediated 5mC oxidation correlates with the DNA methylation level of genomic DNA. Therefore, global DNA methylation levels of human genomic DNA can be determined by quantifying the succinate in 5mC oxidation (Fig. 2). TET2 catalytic domain can be expressed in *Escherichia coli* (*E. coli*), and purified using the Streptag system. The amount of succinate produced by TET-mediated 5mC oxidation in the reaction solution can be measured using an enzymatic succinate quantification kit without any purification. The succinate produced from TET2-mediated reaction is increased with an increasing global DNA methylation level of human genomic DNA. BRET assay requires at least 2.2μg genomic DNA for the quantification of global DNA methylation levels, whereas TET-based assay quantifies global DNA methylation levels using as little as 100 ng genomic DNA. This chapter describes the details of the preparation of TET2 using *E. coli* and the quantification of succinate for the measurement of DNA methylation levels of genomic DNA.

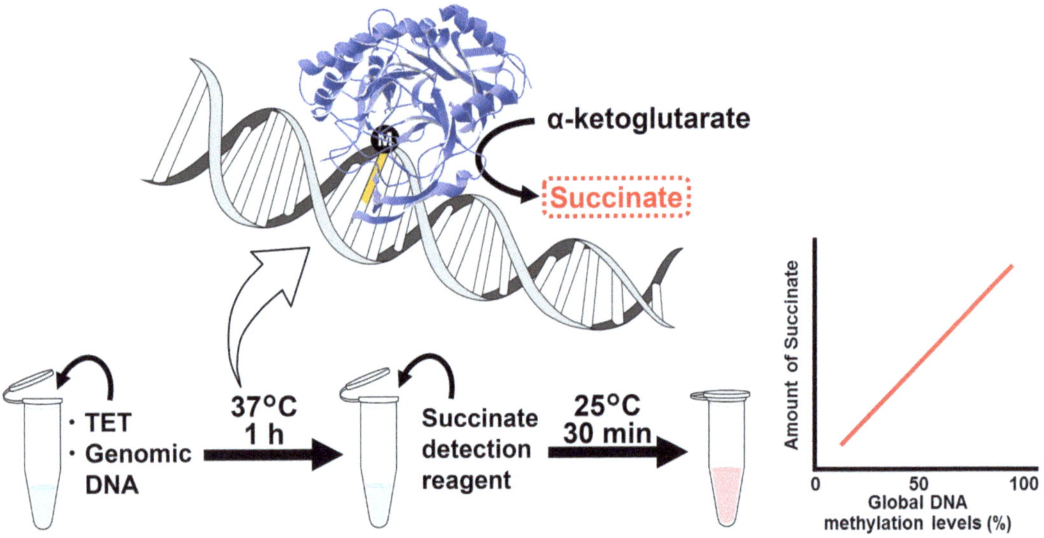

Fig. 2 Quantification of global DNA methylation levels using TET-mediated reaction. The amount of succinate produced from TET-mediated 5mC oxidation is correlated with global DNA methylation levels. This simple assay can quantify global DNA methylation levels using TET without any purification

2 Materials

2.1 Protein Expression

2.1.1 Equipment

1. Incubator (e.g., Double Chamber Incubator INC820, Yamato Scientific).
2. Shaker (e.g., Universal Shaker SHK-U4, IWAKI).
3. Refrigerated Centrifuge (e.g., High Speed Refrigerated Centrifuge Model 7000, KUBOTA).
4. Tabletop centrifuge (e.g., himac CT15RE, HITACHI).
5. Microtube rotator (e.g., MTR103, AS ONE).
6. Variable-speed pump for protein purification (e.g., Variable-Flow Peristaltic Pumps, Fisher Scientific).

2.1.2 Reagents and Consumables

1. *E. coli* strain BL21 (DE3) (e.g., BioDynamics Laboratory).
2. LB medium: 10 mg/mL tryptone, 5 mg/mL yeast extract, 10 mg/mL NaCl, pH 7.2.
3. 50 mg/mL kanamycin.
4. 0.25 M isopropyl β-D-1-thiogalactopyranoside (IPTG).
5. 0.85% NaCl.
6. BugBuster Protein Extraction Reagent (Millipore).
7. 10× PBS buffer: 1.37 M NaCl, 0.02 M KCl, 0.080 M Na_2HPO_4, 0.02 M KH_2PO_4, pH 7.4.
8. NP buffer: 50 mM NaH_2PO_4, 300 mM NaCl, pH 8.0.
9. NPD buffer: 50 mM NaH_2PO_4, 300 mM NaCl, 2.5 mM desthiobiotin, pH 8.0.

10. 0.45 μm filter (e.g., Millipore).
11. Strep-Tactin Superflow column 1 mL (Qiagen).
12. Centrifugal filter for protein concentration (e.g., Amicon® Ultra-15 Centrifugal Filter Unit 30K device, Millipore).
13. Protein concentration determination kit (e.g., Protein Assay BCA Kit, Wako).

2.2 Quantification of Succinate

2.2.1 Equipment

1. Thermal cycler (e.g., T100™ Thermal Cycler, BIO-RAD).
2. Microplate reader (e.g., SpectraMax iD5, Molecular Devices).

2.2.2 Reagents and Consumables

1. DNA purification kit (e.g., DNeasy®, Qiagen).
2. Succinate quantification kit (EnzyChrom Succinate Assay Kit, BioAssay Systems).
3. HEPES buffer: 250 mM HEPES-NaOH, pH 8.0.
4. 1.5 mM $Fe(NH_4)_2(SO_4)_2$.
5. 60 mM ascorbic acid.
6. 30 mM α-ketoglutaric acid.
7. 96-well black plate (e.g., Sanplatec).

3 Methods

3.1 TET2 Expression Vector

Global DNA methylation levels are measured by the amount of succinate produced from TET-mediated 5mC oxidation. TET proteins contain three members (TET1, TET2, and TET3) [48]. TET2 (2002 amino acids) has high enzymatic activity in vitro, especially among TET proteins [49, 50]. Therefore, the human TET2 catalytic domain (1129–1936 amino acids with 1481–1843 replaced with a 15-residue GS linker) with a Streptag at the N-terminal is cloned into pET30c. The TET2 expression vector is available from RIKEN BRC (RDB18377).

3.2 Preparation of TET2

3.2.1 Protein Expression

1. Transform *E. coli* strain BL21 (DE3) by pET30c-Streptag-TET2.
2. Transformants are grown overnight at 37 °C with shaking at 180 rpm in 10 mL LB medium containing 50 μg/mL kanamycin.
3. Add the overnight cultures to 150 mL fresh LB medium containing 50 μg/mL kanamycin, and cultivate at 37 °C with shaking at 180 rpm until the OD_{600} reaches 0.8.
4. Induce the expression by adding IPTG (final concentration of 0.5 mM), and incubate the culture for an additional 18 h at 16 °C with shaking at 180 rpm.

5. Harvest the cells by centrifugation at 2500 × *g* at 4 °C for 10 min using a refrigerated centrifuge.
6. Suspend the pellets in 35 mL 0.85% NaCl.
7. Centrifuge at 2500 × *g* at 4 °C for 10 min using a refrigerated centrifuge, and then remove the supernatant.
8. Repeat **steps 6** and 7 twice.
9. Store the pellets at −80 °C before purification (*see* **Note 1**).

3.2.2 Protein Purification

1. Lyse 1.0 g cell pellets in 5 mL protein extraction solution (500 μL BugBuster Protein Extraction Reagent, 1× PBS), and mix well at room temperature for 15 min using the rotator.
2. Centrifuge at 16,000 × *g* at 4 °C for 20 min, and collect the supernatant as the water-soluble fraction.
3. The water-soluble fraction is diluted to 10 mL in 1× PBS and filtered using a 0.45 μm filter.
4. Inject the filtered fraction to Strep-Tactin Superflow column.
5. Wash the column with 10 mL NP buffer, and then elute TET2 from the column by 10 mL NPD buffer at a flow rate of 1 mL/min. The NP buffer and NPD buffer should be kept at 4 °C during the purification process.
6. Concentrate the 10 mL elution fraction using a centrifugal filter (*see* **Note 2**).
7. Store the concentrated elution fraction with 50% glycerol and 5 mM DTT at −30 °C (*see* **Note 3**).

3.3 Global DNA Methylation Analysis by TET2

1. Purify genomic DNA from targeted cells using a commercial kit or general purification method.
2. All samples to be analyzed should be prepared in triplicate in a total volume of 20 μL, as detailed below.
   (a) 100 ng genomic DNA.
   (b) 50 mM HEPES-NaOH.
   (c) 75 μM $Fe(NH_4)_2(SO_4)_2$.
   (d) 2 mM ascorbic acid.
   (e) 1 mM α-ketoglutaric acid.
   (f) 0.7 μM TET2.
3. Incubate at 37 °C for 1 h in a thermal cycler.
4. Prepare Working Reagent by mixing 85μL Assay Buffer, 1μL Enzyme Mix, 1 μL Cosubstrate, 1 μL PEP, and 1 μL Dye Reagent on ice before immediate use. Add 80 μL of the Working Reagent to 20 μL of each samples.
5. Incubate at 25 °C for 30 min in a thermal cycler.

6. Fluorescence intensity is measured by a microplate reader ($\lambda_{ex} = 530$ nm, $\lambda_{em} = 585$ nm, integration time of 400 ms, 37 °C).
7. Measured data are calculated by subtracting the fluorescence intensity in the presence of TET2 and Working Reagent without genomic DNA. In the case of HeLa genomic DNA of each methylation level (31.8%, 37.9%, 45.4%, and 67.2%), the calculated intensities linearly increase with increasing global DNA methylation levels (*see* **Note 4**).

4 Notes

1. Approximately 1.0 g pellet would be obtained from 150 mL bacterial culture.
2. To concentrate the purified protein, the Amicon® Ultra-15 Centrifugal Filter Unit 30K Device is used in this study. After rinsing the centrifugal filter with ultrapure water, 10 mL purified protein is added to the filter. The centrifugal filter is centrifuged at 4000 × *g* at 4 °C for 20 min.
3. The concentration of the concentrated elution fraction is measured by a protein determination kit. In this protocol, approximately 5.6 μM protein is obtained.
4. In a previous study [47], the genomic DNA of each methylation level was prepared using HeLa cells treated with the DNA methyltransferase inhibitor 5-Aza-2′-deoxycytidine. HeLa cells (5.0×10^5 cells) were seeded into 10 cm dishes. 5-Aza-2′-deoxycytidine was added to each dish (final concentration of 0, 0.1, 1.0, or 10 μM). The medium was changed by a fresh medium without 5-Aza-2′-deoxycytidine after 24 h incubation. After an additional incubation of 3 days, the cells were harvested at 1000 × *g* for 5 min. Genomic DNA was extracted from these cells using a commercial kit (DNeasy®). The methylation levels were determined by the combined bisulfite restriction analysis (COBRA) against the LINE1 region. Sodium bisulfite treatment was performed for genomic DNA (1 μg) using a bisulfite conversion kit (EpiTect Bisulfite Kit, Qiagen). Polymerase chain reaction (PCR) was carried out in a volume of 50 μL containing 250 ng bisulfite-treated genomic DNA, 10× ExTaq buffer (Takara), 1 mM dNTP mix, 1 unit ExTaq HS polymerase (Takara), and 25 pmol of each primer [forward: 5′-G(T/C)GTAAGGGGTTAGGGAGTTTTT-3′ and reverse: 5′-AAC(A/G)TAAAACCCTCC(A/G)AAC-CAAATATAAA-3′] to amplify the LINE1 region. The thermocycling conditions were as follows: 95 °C for 5 min followed by 35 cycles of 95 °C for 10 s, 55 °C for 30 s, and 72 °C for 1 min.

After purification of PCR products using a DNA purification kit (Wizard SV Gel and PCR Clean-Up System, Promega), DNA (200 ng) was treated with 2 units of the restriction enzyme *Taq*$^{\alpha}$I at 65 °C for 1 h. The products were subjected to 15% polyacrylamide gel electrophoresis in 1× Tris-borate-EDTA (TBE) buffer and stained with ethidium bromide. CpG methylation levels in the LINE1 region were determined by calculating the ratio of the band intensity of the digested product to the total amplified PCR product. These band intensities were measured by computer-assisted image analysis software (ImageJ).

Acknowledgments

This work was supported by the Precise Measurement Technology Promotion Foundation.

References

1. Cash HL, Tao L, Yuan JM et al (2012) LINE-1 hypomethylation is associated with bladder cancer risk among non-smoking Chinese. Int J Cancer 130:1151–1159
2. Varriale A (2014) DNA methylation, epigenetics, and evolution in vertebrates: facts and challenges. Int J Evol Biol 2014:475981
3. Jones PA, Takai D (2001) The role of DNA methylation in mammalian epigenetics. Science 293:1068–1070
4. Jaenisch R, Bird A (2003) Epigenetic regulation of gene expression: how the genome integrates intrinsic and environmental signals. Nat Genet 33:245–254
5. Brueckner B, Boy RG, Siedlecki P et al (2005) Epigenetic reactivation of tumor suppressor genes by a novel small-molecule inhibitor of human DNA methyltransferases. Cancer Res 65:6305–6311
6. Bostick M, Kim JK, Estève PO et al (2007) UHRF1 plays a role in maintaining DNA methylation in mammalian cells. Science 317:1760–1764
7. Sharif J, Muto M, Takebayashi S et al (2007) The SRA protein Np95 mediates epigenetic inheritance by recruiting Dnmt1 to methylated DNA. Nature 450:908–912
8. Fang J, Cheng J, Wang J et al (2016) Hemimethylated DNA opens a closed conformation of UHRF1 to facilitate its histone recognition. Nat Commun 7:11197
9. Tahiliani M, Koh KP, Shen Y et al (2011) Conversion of 5-methylcytosine to 5-hydroxymethylcytosine in mammalian DNA by MLL partner TET1. Science 324:930–935
10. Ito S, Shen L, Wu SC et al (2011) Tet proteins can convert 5-methylcytosine to 5-formylcytosine and 5-carboxylcytosine. Science 333:1300–1303
11. Iyer LM, Tahiliani M, Rao A et al (2009) Prediction of novel families of enzymes involved in oxidative and other complex modifications of bases in nucleic acids. Cell Cycle 8:1698–1710
12. Kohli RM, Zhang Y (2013) TET enzymes, TDG and the dynamics of DNA demethylation. Nature 502:472–479
13. Lu X, Zhao BS, He C (2015) TET family proteins: oxidation activity, interacting molecules, and functions in diseases. Chem Rev 115:2225–2239
14. An J, Rao A, Ko M (2017) TET family dioxygenases and DNA demethylation in stem cells and cancers. Exp Mol Med 49:e323
15. Long HK, Blackledge NP, Klose RJ (2013) ZF-CxxC domain-containing proteins, CpG islands and the chromatin connection. Biochem Soc Trans 41:727–740
16. Ko M, An J, Bandukwala HS et al (2013) Modulation of TET2 expression and 5-methylcytosine oxidation by the CXXC domain protein IDAX. Nature 497:122–126
17. Kuiper C, Vissers MCM (2014) Ascorbate as a co-factor for Fe- and 2-oxoglutarate dependent dioxygenases: physiological activity in tumor growth and progression. Front Oncol 4:359

18. Young JI, Zuchner S, Wang G (2015) Regulation of the epigenome by vitamin C. Annu Rev Nutr 35:545–564
19. Blaschke K, Ebata KT, Karimi MM et al (2013) Vitamin C induces Tet-dependent DNA demethylation and a blastocyst-like state in ES cells. Nature 500:222–226
20. Yin R, Mao SQ, Zhao B et al (2013) Ascorbic acid enhances Tet-mediated 5-methylcytosine oxidation and promotes DNA demethylation in mammals. J Am Chem Soc 135:10396–10403
21. Xu Y, Liu SY, Li J et al (2018) Real-time sensing of TET2-mediated DNA demethylation *in vitro* by metal-organic framework-based oxygen sensor for mechanism analysis and stem-cell behavior prediction. Anal Chem 90:9330–9337
22. Rodriguez J, Frigola J, Vendrell E et al (2006) Chromosomal instability correlates with genome-wide DNA demethylation in human primary colorectal cancers. Cancer Res 66:8462–9468
23. Tan AC, Jimeno A, Lin SH et al (2009) Characterizing DNA methylation patterns in pancreatic cancer genome. Mol Oncol 3:425–438
24. Barchitta M, Quattrocchi A, Maugeri A et al (2014) LINE-1 hypomethylation in blood and tissue samples as an epigenetic marker for cancer risk: a systematic review and meta-analysis. PLoS One 9:e109478
25. Chen RZ, Pettersson U, Beardet C et al (1998) DNA hypomethylation leads to elevated mutation rates. Nature 395:89–93
26. Chalitchagorn K, Shuangshoti S, Hourpai N et al (2004) Distinctive pattern of LINE-1 methylation level in normal tissues and the association with carcinogenesis. Oncogene 23:8841–8846
27. Ai S, Shen L, Guo J et al (2012) DNA methylation as a biomarker for neuropsychiatric diseases. Int J Neurosci 122:165–176
28. Rao JS, Keleshian VL, Klein S et al (2012) Epigenetic modifications in frontal cortex from Alzheimer's disease and bipolar disorder patients. Transl Psychiatry 2:e132
29. Coppieters N, Dieriks BV, Lill C et al (2014) Global changes in DNA methylation and hydroxymethylation in Alzheimer's disease human brain. Neurobiol Aging 35:1334–1344
30. Sherwani SI, Khan HA (2015) Role of 5-hydroxymethylcytosine in neurodegeneration. Gene 570:17–24
31. Masliah E, Dumaop W, Galasko D et al (2013) Distinctive patterns of DNA methylation associated with Parkinson disease: identification of concordant epigenetic changes in brain and peripheral blood leukocytes. Epigenetics 8:1030–1038
32. Castillo S, Muñoz P, Behrens MI et al (2017) On the role of mining exposure in epigenetic effects in Parkinson's disease. Neurotox Res 32:172–174
33. Xiong Z, Laird PW (1997) COBRA: a sensitive and quantitative DNA methylation assay. Nucleic Acids Res 25:2532–2534
34. Yang AS, Estécio MRH, Doshi K et al (2004) A simple method for estimating global DNA methylation using bisulfite PCR of repetitive DNA elements. Nucleic Acids Res 32:e38
35. Baba Y, Huttenhower C, Nosho K et al (2010) Epigenomic diversity of colorectal cancer indicated by LINE-1 methylation in a database of 869 tumors. Mol Cancer 9:125
36. Kim JS, Chung WC, Lee KM et al (2012) Association between genetic instability and Helicobacter pylori infection in gastric epithelial dysplasia. Gastroenterol Res Pract 2012:360929
37. Armstrong KM, Bermingham EN, Bassett SA et al (2011) Global DNA methylation measurement by HPLC using low amounts of DNA. Biotechnol J 6:113–117
38. Kremer D, Metzger S, Kolb-Bachofen V et al (2012) Quantitative measurement of genome-wide DNA methylation by a reliable and cost-efficient enzyme-linked immunosorbent assay technique. Anal Biochem 422:74–78
39. Balaghi M, Wagner C (1993) DNA methylation in folate deficiency: use of CpG methylase. Biochem Biophys Res Commun 193:1184–1190
40. Rampersaud GC, Kauwell GP, Hutson AD et al (2000) Genomic DNA methylation decreases in response to moderate folate depletion in elderly women. Am J Clin Nutr 72:998–1003
41. Pilsner JR, Liu X, Ahsan H et al (2007) Genomic methylation of peripheral blood leukocyte DNA: influences of arsenic and folate in Bangladeshi adults. Am J Clin Nutr 86:1179–1186
42. Yoshida W, Baba Y, Karube I (2016) Global DNA methylation detection system using MBD-fused luciferase based on bioluminescence resonance energy transfer assay. Anal Chem 88:9264–9268
43. Baba Y, Karube I, Yoshida W (2019) Global DNA methylation level monitoring by methyl-CpG binding domain-fused luciferase. Anal Lett 52:754–760
44. Yoshida W, Baba Y, Banzawa K et al (2017) A quantitative homogeneous assay for global DNA methylation levels using CpG-binding domain- and methyl-CpG-binding domain-fused luciferase. Anal Chim Acta 990:168–173

45. Taka N, Karube I, Yoshida W (2019) Direct detection of hemi-methylated DNA by SRA-fused luciferase based on bioluminescence resonance energy transfer. Anal Lett 52:1258–1267
46. Baba Y, Yamamoto K, Yoshida W (2019) Multicolor bioluminescence resonance energy transfer assay for quantification of global DNA methylation. Anal Bioanal Chem 411:4765–4773
47. Taka N, Yoshida W (2020) Quantification of global DNA methylation level using 5-methylcytosine dioxygenase. Anal Bioanal Chem 412:5299–5305
48. Rasmussen KD, Helin K (2016) Role of TET enzymes in DNA methylation, development, and cancer. Genes Dev 30:733–750
49. Hu L, Li Z, Cheng J et al (2013) Crystal structure of TET2-DNA complex: insight into TET-mediated 5mC oxidation. Cell 155:1545–1555
50. Hu L, Lu J, Cheng J et al (2015) Structural insight into substrate preference for TET-mediated oxidation. Nature 527:118–122

Part II

Detection of 6mA

Chapter 10

Metabolically Generated Stable Isotope for Identification of DNA *N*⁶-Methyladenine Origin in Cultured Mammalian Cells

Baodong Liu and Hailin Wang

Abstract

DNA N^6-methyladenine (6mA) is a prevalent epigenetic modification in prokaryotes and protists and a potential epigenetic marker in invertebrates, such as *Drosophila melanogaster*, *Chlamydomonas*, and *C. elegans*. However, the existence of 6mA in vertebrates (including mammals) remains elusive. This can be caused at least by two factors. One is that the abundance of 6mA is too low to be detected (generally far less than one 6mA per million deoxyadenine). Essentially, the accurate measurements of rare DNA 6mA in mammals are heavily dependent on ultrasensitive liquid-chromatography-coupled mass spectrometry, which requires skillful operators. The second is the widespread of prokaryotes (e.g., mycoplasma), which contain abundant DNA 6mA with at least five orders of magnitudes higher than mammals. The widespread prokaryotic DNA would contaminate mammalian DNA, resulting in false positive measurements. To solve this problem, here, we describe a new heavy stable isotope-labeled deoxyadenine tracing strategy and a detailed protocol for accurate and sensitive measurement of DNA 6mA in cultured mammalian cells.

Keywords DNA N^6-methyladenine, Isotope labeling, Mass spectrometry, Mammalian DNA, Bacterial DNA

1 Introduction

DNA 6mA is the product with an exocyclic methylated NH2 group. This DNA modification is abundant in genomes of prokaryotes, such as bacteria, archaea, mycoplasma, and fungi. Since it was found in genomes of higher eukaryotes, such as *Drosophila* melanogaster [1], *Chlamydomonas* [2], and *C. elegans* [3] in 2015, DNA 6mA has been detected in mouse embryonic stem cell [4], human [5], zebrafish and pig [6], rats and xenopus laevis [7–10], etc. 6mA has been shown as a probably heritable epigenetic mark in *C. elegans* [3] and involves activating or suppressing gene and transposon expression [1, 2, 4, 11, 12].

A number of methods have been developed for detection of 6mA at global levels, such as high performance liquid chromatography-mass spectrometry (HPLC-MS) [1, 9, 13], dot

Bi-Feng Yuan (ed.), *DNA Modification Detection Methods* , Springer Protocols Handbooks,
https://doi.org/10.1007/978-1-0716-1229-3_10,

blot, immunofluorescence imaging [6, 14], and for sequencing of 6mA to elucidate its location and distribution, e.g., antibody-based immunoprecipitation-coupled genome-wide sequencing and single-molecule real-time sequencing (SMRT-seq) [1–4, 6]. However, up to date, most methods have their disadvantages. For example, antibody dependent immunoassay techniques cannot effectively distinguish 6mA with RNA m6A. So, it is required to completely remove RNA away from the tested samples. Otherwise, the enriched RNA m6A would generate false positive for dot blot and for DNA immunoprecipitation-based next-generation sequencing. LC-MS/MS has excellent sensitivity and very low detection limit of 6mA and can substantially distinguish 6mA from m6A, 1mA and 3mA by their difference in molecular mass and LC retention. Both antibody dependent assay and LC-MS method cannot provide sequence and distribution information of 6mA in genes and encounter prokaryotic DNA contamination. DNA sequencing technology can tell the 6mA sites in genomes and can partly deduct prokaryotic DNA background. Generally, two or more techniques were used to verify the true level of 6mA in genome with rigorously designed experiments. However, all of these methods above-mentioned have their limitations on the measurement of the vanishingly low abundance of 6mA in mammalian genome (mouse embryonic stem cells (0.0001–0.0005%) [4]), accompanying with abundant prokaryotic DNA contamination. It should be noted that whether there is detectable 6mA in genome of mouse embryonic stem cells is debatable. In 2016, Xiao's group used LC-MS/MS technology and showed the presence of 6mA in mouse embryonic stem cells. Then, they adopted SMRT technology and identified the 6mA enriched in young full length line-1 [4]. However, 2017, Thomas Carell et al. reported no 6mA was detected in mouse embryonic stem cells using UHPLC-QQQ-MS/MS [15]. In 2018, Fang's group exploited SMRT-seq and showed that there was no detectable 6mA in young full length line-1 of human and mice by subtracting background signal [16].

To overcome the contamination of prokaryotic DNA, we developed a metabolic stable isotope tracing for detection of 6mA in culturing mammalian cells, as shown in Fig. 1.

We showed that stable isotope-labeled $^{15}N_5$-dA is incorporated into genomic DNA mainly in the form of $^{15}N_4$-dA by exocyclic deamination of dA [17]. This process relies on adenosine deaminase and purine salvage pathway. Compared to the unlabeled natural dA, $^{15}N_5$- and $^{15}N_4$- labeled carry additional 5.0 and 4.0 amu, respectively. If it is a methylated form of dA, 6mA can be formed by activity of DNA adenine methyltransferases in vivo. So, if there is activity of DNA adenine methyltransferases in vivo, heavily labeled $^{15}N_4$-6mA or $^{15}N_5$-6mA should be detected in $^{15}N_5$-dA-labeled genome, too. The DNA carried by bacteria only contains natural isotope, there should be negligible $^{15}N_4$-6mA and $^{15}N_5$-6mA

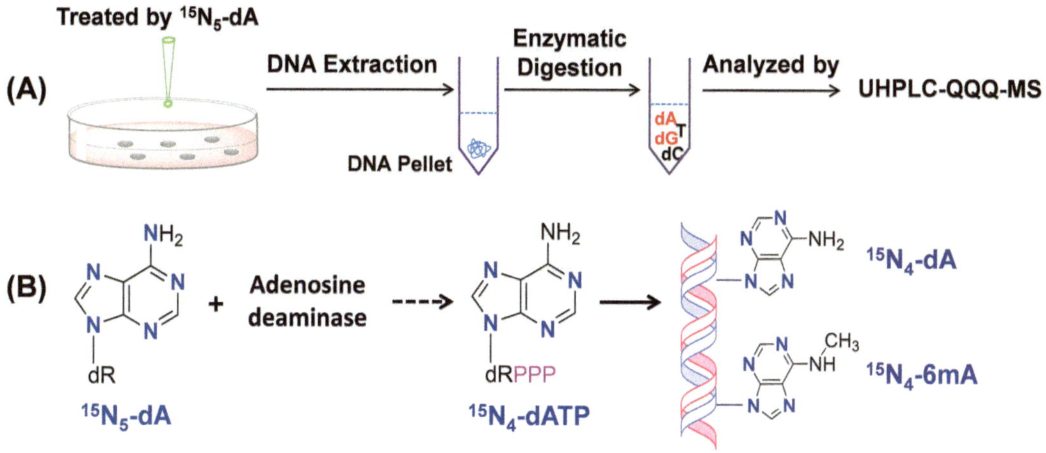

Fig. 1 Diagram of $^{15}N_5$-dA tracing for detection of endogenous 6mA in genomes of cultured cells. (**a**) Flow chart of tracing experiment; (**b**) Schematic of incorporation of deoxyadenine into the genome by salvage pathway in culturing mammalian cells

relative to the unlabeled 6mA. Based on the principles described above, endogenous 6mA originated from the methylation activity of mammalian cells itself can be verified even in the presence of prokaryotic DNA contamination. A detailed protocol used stable isotope-labeled deoxyadenine tracing for detection of 6mA is given as follows.

2 Materials

2.1 Cell Culture and Cell Proliferation Detection

1. 10× Dulbecco's phosphate buffered saline (sterile PBS, Sigma-Aldrich, Cat. #59331C, 2 g/L potassium chloride, 2 g/L potassium phosphate monobasic anhydrous, 80 g/L sodium chloride, 21.716 g/L sodium phosphate dibasic heptahydrate, without calcium and magnesium).
2. Trypsin (Hyclone, Cat. #SH30042.01, with 2.5 g porcine trypsin in HBSS with 0.2 g/L ethylene diamine tetraacetic acid, without calcium, magnesium).
3. 70% ethanol.
4. Six-well cell culture cluster, 96-well cell culture cluster, and 60 mm culture dishes (Corning, Cat. #3516, Cat. #3599, Cat. #430166).
5. 5 mL sterility tubes (Sangon, Cat. #F611888).
6. Dulbecco's modified Eagle's medium (DMEM, Hyclone, Cat. #SH30022.01, without sodium pyruvate) containing 4 mM L-glutamine, 4.5 g/L glucose.
7. Fetal bovine serum (FBS, Biological Industries, Lot #1421135).

8. 100× Penicillin-Streptomycin solution: 10,000 units/mL Penicillin, 10,000μg/mL Streptomycin in 0.85% NaCl (Hyclone, Cat. #SV30010).
9. 2′-Deoxyadenosine monohydrate (purity: ≥99%, Sigma, Cat. #D7400) is and $^{15}N_5$-dA (purity: 96–98%, Cambridge Isotope Laboratories, Cat. #NLM-3895). Powder of dA and $^{15}N_5$-dA was dissolved in sterile water to the final concentration of 10 mM, separately, and stored at −20 °C.
10. CellTiter 96® AQueous One Solution Cell Proliferation Assay (Promega, Cat. #G3580).

2.2 Mycoplasma Detection

1. PCR mycoplasma detection kit (HuaBio, Cat. #K0103).
2. Agarose powder (Amresco, Cat. #4-2070-EA).
3. Gel Red stain solution (BioRoYee, Cat. #15-3017).
4. DNA Marker I (Biomed, Cat. #MD107-01).

2.3 DNA Extraction

1. Genomic DNA extraction kit (Promega, Cat. #A1120).
2. Isopropanol.
3. 1 mg/mL RNAse A (Tiangen, Cat. #RT405-02).

2.4 DNA Digestion and Ultrafiltration

1. 1000 U/mL Deoxyribonuclease I (DNase I, New England Biolabs, Cat. #M0303S).
2. Quick calf intestinal alkaline phosphatase (CIP, New England Biolabs, Cat. #M0525S).
3. Crotalus adamanteus venom phosphodiesterase I (SVP, Worthington Biochemical Corporation, Cat. #LS003926).
4. Ultrafiltration centrifugal tube (molecular weight cutoff 3000 Da, Pall, Cat. # MCP003C46).

2.5 UHPLC-MS/MS Analysis

1. Zorbax Eclipse Plus C18 column (50 mm × 2.1 mm I.D., 1.8μm particle size, Agilent Technologies, Cat. #959757-902 K).
2. Ultrapure water used for LC-MS system was prepared by ELGA PureLab-Ultra purifier (High Wycombe).
3. LC-MS grade methanol (Fluka, Cat. #34966).
4. LC-MS grade Formic acid (Sigma-Aldrich, Cat. #5330020050).
5. 0.5 mL sample bottles (MN, Cat. #702009).

3 Methods

3.1 Cell Culture of HEK293T Cells

HEK293T cells were seeded in 60 mm culture dishes and grown in 3 mL DMEM medium, which is supplemented with 10% FBS,

100μg/mL streptomycin, and 100 U/mL penicillin. The dishes were incubated at 37 °C in a humidified incubator with 5% CO_2. When cell density of HEK293T cells reached 80 ~ 90%, an appropriate number of cells were seeded in 96-well plate for MTS assay or 6-well plate for $^{15}N_5$-dA tracing assay.

3.2 Detection of Mycoplasma in Cell Culture

Mycoplasma is a common and unobservable contamination in cell culture. It was reported that about 5–35% cultured cells are contaminated with mycoplasma [18]. The *Mycoplasma fermentans*, *M. orale*, *M. arginini*, *M. hyorhinis*, and *Acholeplasma laidlawii* are the most common species in the laboratory-cultured cells. There are several mycoplasma detection methods, such as broth and agar culture, direct DNA staining, ELISA (Enzyme linked immunosorbent assay), immunostaining, PCR, etc. Broth and agar culture are the golden standard of mycoplasma detection; however, it is slow and need skillful operation. PCR is a frequently used, fast, and reliable method for the detection of mycoplasma. To detect potential mycoplasma contamination, primers that have been confirmed by previous reports or commercial PCR mycoplasma detection kits are used in our lab. The PCR products with different sizes are analyzed by agarose gel electrophoresis. Note: positive and negative controls are required.

3.3 The Toxicity of Deoxyadenine for Cultured Cells

It is necessary to determine cell viability for different cell lines after exposure to heavy stable isotope-labeled dA, because it may have different cytotoxicity upon cell types. There are several methods used for measuring cell viability, such as MTT (3-(4,5-dimethlthiazol-2-yl)-2,5-diphenyltetrazolium bromide) assay, MTS (3-(4,5-dimethylthiazol-2-yl)-5-(3-carboxymethoxyphenyl)-2-(4-sulfophenyl)-2H-tetrazolium) assay, Ki-67 cell proliferation assay, etc. To evaluate the toxicity of dA on the cell viability of HEK293T cells, MTS assay is performed according to the manufacturer's instruction.

1. HEK293T cells (10,000 cells/well) are seeded in 96-well plate and added with 100μL medium containing 20, 40, 60, 80, 100, 200, 300, and 400μM unlabeled dA. Untreated group (no dA) is used as control. Cultured medium without cells is used as the background control. Three replicates are set for each concentration point.
2. The cells cultured in 96-well plates are incubated for 24 h at 37 °C.
3. 20μL of MTS solution is added to each well, and then the plate is incubated for another 2 h at 37 °C in dark.
4. The plates are detected at 492 nm by an automatic microplate reader (Thermo Fisher Scientific).
5. Cell viability of treatment group is calculated by the equation:

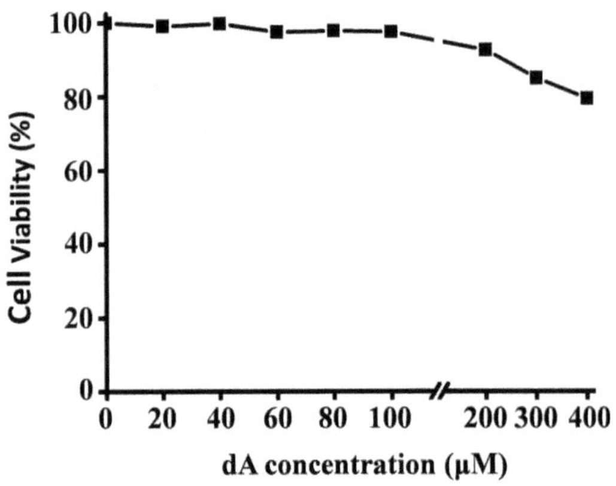

Fig. 2 Viability of HEK293T cells treated with different concentrations of dA. (Reproduced from ref. [17] with permission from American Chemical Society)

$$\frac{\text{Mean OD(dA-treated)} - \text{Mean OD(background control)}}{\text{Mean OD(non-dA-treated)} - \text{Mean OD(background control)}} \times 100\%$$

The cell viability of no-dA-treated group was set as 100%. The result is shown in Fig. 2.

3.4 Cell Treatment Using Appropriate Concentration of $^{15}N_5$-dA

Many factors affect the stable isotope labeling ratio of dA, e.g., $^{15}N_5$-dA concentration, cell density, cell types, etc. The highest concentration of $^{15}N_5$-dA without affecting cell growth is the optimum to achieve a high labeling ratio. However, even low labeling ratio of dA can meet the requirement of the experiments for identification of 6mA origin.

1. 9×10^5 HEK293T cells/well are seeded in 6-well plates and cultured with 2 mL medium with 20μM $^{15}N_5$-dA. Cells grown in the medium without the treatment of $^{15}N_5$-dA is used as negative control.
2. The 6-well plate is incubated for 24 h at 37 °C. The cells are harvested for DNA extraction and mycoplasma detection.
3. To improve the labeling ratio of dA, if necessary, the cells can be passaged every 2 day and cultured in fresh medium containing 20μM $^{15}N_5$-dA as mentioned above.

3.5 Genomic DNA Preparation

Cells are harvested and genomic DNA is extracted using Genomic DNA Purification Kit following the manufacturer's instructions. The general experimental procedure is listed as follows.

1. Medium is removed from 6-well plate by sterile pipette.
2. Then, cells are gently washed with 1 mL PBS twice.

3. PBS is removed and 500μL trypsin per well is added. Then, the plates are incubated at 37 °C for 2 min.
4. 1 mL complete medium per well is added to terminate the digestion. The resuspension solution of cells is transferred to a new 2 mL tube.
5. Cells are harvested by centrifugation (100g, 5 min).
6. Cell pellets are gently and adequately dispersed by pipette with 1.0 mL PBS.
7. Cells are collected by centrifugation again (100g, 5 min).
8. The PBS is carefully removed and 600μL Nuclei Lysis Solution is added to fully lysed cells.
9. Then, 5μL of RNase A solution is added and fully mixed and the tubes are incubated at 37 °C for 30 min.
10. 200μL protein precipitation solution is added and mixed by inversion the tube several times.
11. Centrifuge for 20 min at 13,000 g at 4 °C. Then, transfer the supernatant to a new 1.5 mL centrifuge tube.
12. Add 600μL of isopropanol, and acutely invert the tube several times. Usually, white flocculent precipitate can be seen. That is DNA. Gently wash the DNA pellet twice with 70% ethanol.
13. DNA is dried in clean bench and resuspended in 40μL ultrapure water. The quality of DNA is evaluated at the absorbance at 260 nm and 280 nm by Nanodrop 2000 (Thermo Fisher Scientific).

3.6 DNA Digestion and Sample Handling for UHPLC-MS/MS Analysis

DNA is digested with as previously described [19].

1. 5μL of 100 mM Tris-HCl is added to 5μg DNA in a tube.
2. 0.25 U DNase I, 0.25 U SVP, and 0.5 U CIP are added.
3. The total volume of each sample is adjusted to 50 μL with water.
4. The tubes are incubated at 37 °C for 12–16 h.
5. Then, the digestion solution is transferred into an ultrafiltration centrifugal tube (molecular weight cutoff of 3 KDa) and centrifuged at 12,000 g for 30 min. Note, this step is to remove macromolecules and prevent from blocking chromatographic column before the MS analysis system.
6. Filtered samples are transferred to sample vials for UHPLC-MS/MS analysis.

Hydrolysate obtained from the DNA digestion contains four main deoxyribonucleosides (dA, dG, dC, dT) and minor or rare modified deoxyribonucleosides (5mC, 5hmC, etc.). Also, minute amounts of hydrolysis products from other biomacromolecule (RNA and

3.7 UHPLC-MS/MS Analysis of $^{15}N_5$-6mA, $^{15}N_4$-6mA, Unlabled 6mA, $^{15}N_5$-dA, $^{15}N_4$-dA, and Unlabled dA

protein) are unavoidable. So, it is necessary to separate the various compounds by reversed-phase liquid chromatography prior to mass spectrometry detection. Note, heavy stable isotope-labeled deoxyribonucleotides and unlabeled deoxyribonucleotides co-elute because they are hard to be separated by chromatography, although they can be detected by mass spectrometry depending on their molecular weight difference. Several analytical methods for detection of deoxyribonucleosides have been developed in our laboratory [1, 17, 20–22] and one of them is shown as follows.

1. UHPLC conditions [17]. Agilent 1290 series pump equipped with an autosampler and chromatographic column thermostat is used. The temperature of autosampler and column thermostat are set at 4 °C and 30 °C, respectively. Zorbax Eclipse Plus C18 column (50 mm × 2.1 mm I.D., 1.8μm particle size) with C18 cartridge guard-column is used to separate nucleosides. Mobile phase A is water and mobile phase B is methanol. Four main deoxyribonucleosides are separated under isocratic elution (95% solvent A and 5% solvent B). The elution of 6mA is performed with gradient elution: 0.0–2.0 min (5% solvent B), 2.1–4.3 min (from 5 to 20% solvent B), 4.3–6.0 min (20% solvent B), 6.1–9.0 min (100% solvent B), 9.1–13.0 min (5% solvent B for equilibrating column). The flow is 0.3 mL/min and 5μL of sample is used for each injection. The analysis is repeated at least three times for each sample.
2. Mass spectrometry conditions [17]. Agilent 6410B triple quadrupole mass spectrometer is used to analyze the mass difference of isotope labeling nucleosides. The capillary voltage is set at 3500 V. The flow of nitrogen drying gas is 9 L/min. The ion source temperature is set at 300 °C. The collision energy is set at 5 eV for dA and 15 eV for 6mA. The positive ion mode and multiple reaction monitoring (MRM) are used for selective detection: 6mA (266.1→150.1), $^{15}N_4$-6mA (270.1→154.1), $^{15}N_5$-6mA (271.1→155.1), dA (252.1→136.1), $^{15}N_4$-dA (256.1→140.1), and $^{15}N_5$-dA (257.1→141.1).
3. Data analysis. The quantitative analysis of dA and 6mA is calibrated with external standard curve. Peak area of total dA equals to the sum of unlabeled dA, $^{15}N_4$-dA and $^{15}N_5$-dA. Also, the amount of total 6mA equals to the sum of unlabeled 6mA, $^{15}N_4$-6mA, and $^{15}N_5$-6mA. Note, ion pair of 257.1→141.1 stands for not only $^{15}N_5$-dA, but also the stable isotope of $^{15}N_4$-dA (m/z + 1). So, true peak area of $^{15}N_5$-dA equals to the peak area of 257.1→141.1 minus the theoretical peak area of the plus one isotope of $^{15}N_4$-dA. The later can be calculated by Isotope Distribution Calculator (Agilent software package). By the same way, stable isotope-labeled 6mA can be

detected and estimated. Essentially, by subtracting the natural plus one isotope of $^{15}N_4$-dA, the $^{15}N_5$-dA becomes undetectable in HEK293T cells [17] and mouse embryonic stem cells (unpublished data) cultured in our lab. Human adenosine deaminase was proved to play the key role in exocyclic deamination of $^{15}N_5$-dA [17]. However, the unlabeled 6mA, $^{15}N_4$-6mA, and $^{15}N_5$-6mA are undetectable in genomic DNA of HEK293T cells. While, the abundance of unlabeled 6mA, $^{15}N_5$-dA and $^{15}N_5$-6mA are detected in mycoplasma positive HEK293T cells, probably because of deficient activity of mycoplasma adenosine deaminase.

4 Notes

1. We use the liquid chromatography/mass spectrometry (LC/MS) system (Agilent) to measure 6mA, $^{15}N_4$-6mA, and $^{15}N_5$-6mA. However, other types of tandem mass spectrometers are probably applicable.
2. Other precursors of deoxyadenine or methyl donor of 6mA, such as adenosine and S-adenosine methionine, are also available to obtain heavy stable isotope-labeled 6mA in vivo.
3. There is no detectable $^{15}N_5$-labeled-dA in genome of HEK293T cells treated with $^{15}N_5$-dA, because of the strong activity of human adenosine deaminase. Conversely, if activity of adenosine deaminase is inhibited, there should be considerable $^{15}N_5$-dA present in genomic DNA.
4. Frequently assessing and controlling microbial contamination (such as mycoplasma) in cell cultures is necessary.

Acknowledgments

The authors thank the National Natural Science Foundation of China (91743201) and State Key Laboratory of Environmental Chemistry and Ecotoxicology, Research Center for Eco-Environmental Sciences, Chinese Academy of Sciences (KF2017-10) for financial support.

References

1. Zhang G, Huang H, Liu D, Cheng Y, Liu X, Zhang W, Yin R, Zhang D, Zhang P, Liu J, Li C, Liu B, Luo Y, Zhu Y, Zhang N, He S, He C, Wang H, Chen D (2015) N^6-methyladenine DNA modification in Drosophila. Cell 161(4):893–906
2. Fu Y, Luo GZ, Chen K, Deng X, Yu M, Han D, Hao Z, Liu J, Lu X, Dore LC, Weng X, Ji Q, Mets L, He C (2015) N^6-methyldeoxyadenosine marks active transcription start sites in chlamydomonas. Cell 161(4):879–892
3. Greer EL, Blanco MA, Gu L, Sendinc E, Liu J, Aristizabal-Corrales D, Hsu CH, Aravind L,

He C, Shi Y (2015) DNA methylation on N^6-adenine in C. elegans. Cell 161(4):868–878

4. Wu TP, Wang T, Seetin MG, Lai Y, Zhu S, Lin K, Liu Y, Byrum SD, Mackintosh SG, Zhong M, Tackett A, Wang G, Hon LS, Fang G, Swenberg JA, Xiao AZ (2016) DNA methylation on N^6-adenine in mammalian embryonic stem cells. Nature 532 (7599):329–333
5. Xiao C-L, Zhu S, He M, Chen D, Zhang Q, Chen Y, Yu G, Liu J, Xie S-Q, Luo F, Liang Z, Wang D-P, Bo X-C, Gu X-F, Wang K, Yan G-R (2018) N^6-methyladenine DNA modification in the human genome. Mol Cell 71 (2):306–318.e307
6. Liu J, Zhu Y, Luo GZ, Wang X, Yue Y, Wang X, Zong X, Chen K, Yin H, Fu Y, Han D, Wang Y, Chen D, He C (2016) Abundant DNA 6mA methylation during early embryogenesis of zebrafish and pig. Nat Commun 7:13052
7. Yao B, Cheng Y, Wang Z, Li Y, Chen L, Huang L, Zhang W, Chen D, Wu H, Tang B, Jin P (2017) DNA N^6-methyladenine is dynamically regulated in the mouse brain following environmental stress. Nat Commun 8(1):1122
8. Wang Y, Wang Y, Sheng Y, Huang J, Chen X, Al-Rasheid KAS, Gao S (2017) A comparative study of genome organization and epigenetic mechanisms in model ciliates, with an emphasis on tetrahymena, paramecium and oxytricha. Eur J Protistol 61(Pt B):376–387
9. Liang D, Wang H, Song W, Xiong X, Zhang X, Hu Z, Guo H, Yang Z, Zhai S, Zhang LH, Ye M, Du Q (2016) The decreased N^6-methyladenine DNA modification in cancer cells. Biochem Biophys Res Commun 480 (1):120–125
10. Koziol MJ, Bradshaw CR, Allen GE, Costa AS, Frezza C, Gurdon JB (2016) Identification of methylated deoxyadenosines in vertebrates reveals diversity in DNA modifications. Nat Struct Mol Biol 23(1):24–30
11. Xiao CL, Zhu S, He M, Chen ZQ, Chen Y, Yu G, Liu J, Xie SQ, Luo F, Liang Z, Wang DP, Bo XC, Gu XF, Wang K, Yan GR (2018) N^6-methyladenine DNA modification in the human genome. Mol Cell 71(2):306–318. e307
12. Yao B, Li Y, Wang Z, Chen L, Poidevin M, Zhang C, Lin L, Wang F, Bao H, Jiao B, Lim J, Cheng Y, Huang L, Phillips BL, Xu T, Duan R, Moberg KH, Wu H, Jin P (2018) Active N^6-methyladenine demethylation by DMAD regulates gene expression by coordinating with polycomb protein in neurons. Mol Cell 71(5):848–857.e846
13. Huang W, Xiong J, Yang Y, Liu S-M, Yuan B-F, Feng Y-Q (2015) Determination of DNA adenine methylation in genomes of mammals and plants by liquid chromatography/mass spectrometry. RSC Adv 5(79):64046–64054
14. Dabe EC, Sanford RS, Kohn AB, Bobkova Y, Moroz LL (2015) DNA methylation in basal metazoans: insights from ctenophores. Integr Comp Biol 55(6):1096–1110
15. Schiffers S, Ebert C, Rahimoff R, Kosmatchev O, Steinbacher J, Bohne AV, Spada F, Michalakis S, Nickelsen J, Muller M, Carell T (2017) Quantitative LC-MS provides no evidence for m6dA or m4dC in the genome of mouse embryonic stem cells and tissues. Angew Chem Int Ed Engl 56 (37):11268–11271
16. Zhu S, Beaulaurier J, Deikus G, Wu TP, Strahl M, Hao Z, Luo G, Gregory JA, Chess A, He C, Xiao A, Sebra R, Schadt EE, Fang G (2018) Mapping and characterizing N^6-methyladenine in eukaryotic genomes using single-molecule real-time sequencing. Genome Res 28(7):1067–1078
17. Liu B, Liu X, Lai W, Wang H (2017) Metabolically generated stable isotope-labeled deoxynucleoside code for tracing DNA N^6-methyladenine in human cells. Anal Chem 89 (11):6202–6209
18. Young L, Sung J, Stacey G, Masters JR (2010) Detection of mycoplasma in cell cultures. Nat Protoc 5(5):929–934
19. Yin R, Liu S, Zhao C, Lu M, Tang MS, Wang H (2013) An ammonium bicarbonate-enhanced stable isotope dilution UHPLC-MS/MS method for sensitive and accurate quantification of acrolein-DNA adducts in human leukocytes. Anal Chem 85(6):3190–3197
20. Liu X, Lai W, Zhang N, Wang H (2018) Predominance of N^6-methyladenine-specific DNA fragments enriched by multiple immunoprecipitation. Anal Chem 90(9):5546–5551
21. Lai W, Lyu C, Wang H (2018) Vertical ultrafiltration-facilitated DNA digestion for rapid and sensitive UHPLC-MS/MS detection of DNA modifications. Anal Chem 90 (11):6859–6866
22. Lai W, Mo J, Yin J, Lyu C, Wang H (2019) Profiling of epigenetic DNA modifications by advanced liquid chromatography-mass spectrometry technologies. Trend Anal Chem 110:173–182

Chapter 11

Determination of N^6-Methyladenine in DNA of Mammals and Plants by Dpn I Digestion Combined with Size-Exclusion Ultrafiltration and Mass Spectrometry Analysis

Qing-Yun Cheng and Bi-Feng Yuan

Abstract

DNA adenine methylation (N^6-methyladenine, 6mA) widely exists in bacteria and regulates various biological processes including gene transcription and restriction-modification system. In recent years, it has been reported that 6mA is also present in genomic DNA of higher eukaryotes. However, the accurate determination of 6mA in DNA of higher eukaryotes is challenging owing to the potential contamination of bacterial DNA that normally contains abundant 6mA. Therefore, unambiguous detection and quantification method is required for the accurate analysis of 6mA in DNA of higher eukaryotes. We developed a method by Dpn I digestion combined with size-exclusion ultrafiltration and mass spectrometry analysis to determinate 6mA in DNA of higher eukaryotes, including mammals and plants. The contamination of potential bacterial DNA can be efficiently removed from the extracted eukaryotic DNA, which enables the accurate and confident determination of 6mA in DNA of higher eukaryotes.

Keywords DNA adenine methylation, N^6-methyladenine, Dpn I digestion, Size-exclusion ultrafiltration, Liquid chromatography-mass spectrometry

1 Introduction

DNA adenine methylation (N^6-methyladenine, 6mA) is a natural modification occurring in DNA of prokaryotes and eukaryotes [1]. 6mA was first reported to be involved in the restriction-modification systems in prokaryotes [2]. It has also been found that 6mA is involved in DNA mismatch repair and gene regulatory processes in *Escherichia coli* [3]. In 2015, 6mA was discovered to be present in genomic DNA of *Chlamydomonas reinhardtii* [4], *Caenorhabditis elegans* [5], and *Drosophila melanogaster* [6].

In addition to the aforementioned living organisms, 6mA was also found to be present in the genomic DNA of mammals and plants [7–9]. Immunoprecipitation-mediated sequencing revealed the existence of 6mA in genomes of human neurons [10]. The SMRT sequencing demonstrated the presence of 6mA in the

Bi-Feng Yuan (ed.), *DNA Modification Detection Methods*, Springer Protocols Handbooks,
https://doi.org/10.1007/978-1-0716-1229-3_11,

genomic DNA of mouse embryonic stem cells [11] and plants [12, 13]. The modification of 6mA is dynamic and reversible during embryogenesis of vertebrates and mammals [14], indicating that 6mA plays regulatory roles in gene expression [15].

On the other side, several studies showed no detectable 6mA existing in genomic DNA of mammals [16, 17]. A recent study reported only small amount of 6mA in genomic DNA of mammals and proposed that the trace level of 6mA was originated from RNA N^6-methyladenine, which is processed via the nucleotide-salvage pathway and misincorporated by DNA polymerases [18]. He et al. [19] reported that 6mA was mainly enriched in mammalian mitochondrial DNA (mtDNA) and the level of 6mA in HepG2 mtDNA was at least 1300-fold higher than that in genomic DNA.

Liquid chromatography-tandem mass spectrometry (LC-MS/MS) has been frequently used to determine nucleic acid modifications due to its high sensitivity and specificity [20–25]. Isolated genomic DNA was typically enzymatically digested to nucleosides followed by LC-MS/MS analysis [26, 27]. Since bacterial DNA contains most abundant 6mA, potential contamination of tested samples from bacterial DNA may cause the vague and discrepant conclusion of the existence of 6mA in genomes of mammals and plants. In addition, tiny amount of bacterial DNA could compromise the true and accurate analysis of 6mA from eukaryotic genomes. Therefore, removing the possible bacterial DNA contamination from eukaryotic DNA is critical for the unambiguous determination of 6mA in higher eukaryotes.

We developed a method via *Dpn* I digestion combined with size-exclusion ultrafiltration and LC-MS/MS to determine 6mA in genomic DNA [7, 28]. Dpn I can selectively cleave G(6mA)TC sites in DNA. Only the methylated adenine (6mA) in the recognition sites will be cleaved by Dpn I, while normal adenine in GATC sites cannot be cleaved. In this respect, bacterial DNA can be cut into short fragments since adenines in bacteria DNA are heavily modified to 6mA. However, the DNA of mammals and plants remains in large-size after Dpn I digestion due to the extremely low abundance of 6mA in genomic DNA of mammals and plants. The small DNA fragments of bacteria then can be removed from large-size genomic DNA of mammals and plants using size-exclusion ultrafiltration (Fig. 1). The developed method enables efficient elimination of bacterial contamination, which can be used for the unambiguous detection and quantification of 6mA in DNA of mammals and plants.

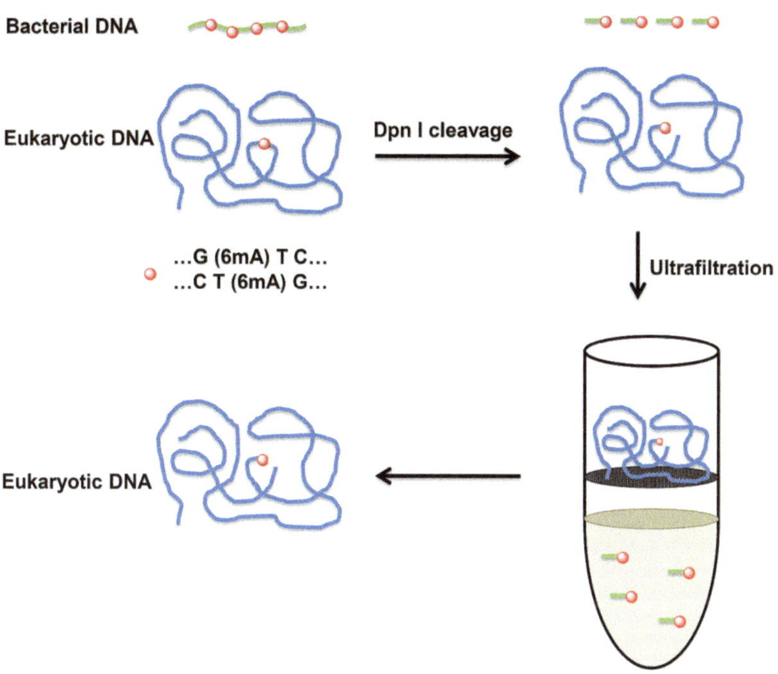

Fig. 1 Schematic diagram for the elimination of potential contamination of bacterial DNA by Dpn I cleavage coupled with size-exclusion ultrafiltration. (Reproduced from ref. [7] with permission from The Royal Society of Chemistry)

2 Materials

Water was prepared from Milli-Q apparatus (Millipore). Chromatographic-grade methanol and acetonitrile (ACN) were used. All other solvents and chemicals are of analytical grade.

2.1 Enzymes

1. S1 nuclease: 180 units/μL.
2. Alkaline phosphatase (CIAP): 30 units/μL.
3. Venom phosphodiesterase I: 0.002 units/μL.
4. DNase I: 5 units/uL.
5. Dpn I restriction enzyme: 20 units/μL.

2.2 Stock Solutions of Standards

1. 2′-Deoxycytidine (dC): 4 mmol/L in water.
2. 2′-Deoxyguanosine (dG): 2 mmol/L in water. (*see* **Note 1**)
3. 2′-Deoxyadenosine (dA): 4 mmol/L in water.
4. Thymidine (T): 2 mmol/L in water.
5. N^6-Methyl-2′-deoxyadenosine (6mA): 4 mmol/L in water.

2.3 Chemicals and Reagents

1. Chloroform.
2. Formic acid.
3. Solvent A: formic acid in water (0.1%, v/v).

4. Solvent B: formic acid in MeOH (0.1%, v/v).
5. 10× enzymatic buffer: 3 uL (500 mM Tris-HCl, 100 mM NaCl, 10 mM $MgCl_2$, 10 mM $ZnSO_4$, pH 7.0).
6. 10× CutSmart buffer (New England Biolabs, Ipswich, WA).
7. DNAiso Reagent (Takara Biotechnology, Dalian, China).
8. E.Z.N.A.® Tissue DNA kit (Omega Bio-Tek Inc., Norcross, GA, USA).
9. E.Z.N.A.® Plant DNA kit (Omega Bio-Tek Inc., Norcross, GA, USA).

2.4 Biological Samples

1. Human leukemia Jurkat-T cells (Jurkat-T) and human embryonic kidney cells (HEK293T) are from the Cell Bank of Type Culture Collection (Chinese Academy of Sciences).
2. Jurkat-T and HEK293T cells were cultured according to previously described method [29, 30]. Add ciprofloxacin (10 μg/mL) in the culture medium to inhibit the possible contamination of *Mycoplasma*.
3. Male Sprague-Dawley rat (4 weeks old) was from the Center for Animal Experiment/ABSL-3 Laboratory of Wuhan University. Sacrifice the rat under anesthesia to collect tissues and store the tissues under −80 °C until use.
4. Plant samples (*Oryza sativa* and *Zea mays*) were cultured according to the previously reported method [31].

2.5 Equipment

1. AB 3200 QTRAP mass spectrometer (Applied Biosystems) coupled with an electrospray ionization source (Turbo Ionspray).
2. Shimadzu LC-20 AD HPLC system with a DGU-20A3 degasser, two LC-20 AD pumps, a SIL-20A autosampler, and a CTO-20 AC thermostated column compartment.
3. Hisep C18-T column (150 mm × 2.1 mm i.d., 5 μm, Weltech Co., Wuhan, China).
4. B-500 spectrophotometer (Metash Instruments Co., Shanghai, China).

3 Methods

3.1 DNA Extraction

1. Extract genomic DNA of Jurkat-T and HEK293T cells using DNAiso Reagent (Takara Biotechnology, Dalian, China) according to the manufacture recommended procedure.
2. Extract genomic DNA of Male Sprague-Dawley rat using E.Z.N.A.® Tissue DNA kit (Omega Bio-Tek Inc., Norcross, GA, USA) according to the manufacture recommended procedure.

3. Isolate genomic DNA of *Oryza sativa* and *Zea mays* using E.Z. N.A.® Plant DNA kit (Omega Bio-Tek Inc., Norcross, GA, USA) according to the manufacture recommended procedure.
4. Determine the concentrations of the isolated DNA with a B-500 spectrophotometer. Obtain the purity and yield of the isolated DNA by measuring the absorbance of DNA at 260 and 280 nm (*see* **Note 2**).

3.2 Dpn I Digestion and Size-Exclusion Ultrafiltration

Digest the extracted genomic DNA by Dpn I restriction enzyme and then remove the possible contamination of bacterial DNA by ultrafiltration (Figs. 1 and 2).

1. 5–10 μg of genomic DNA (in 41μL of H_2O), 5 μL of 10× cutSmart buffer (New England Biolabs, Ipswich, WA), 80 units of Dpn I restriction enzyme (4 μL, New England Biolabs, Ipswich, WA) are gently mixed well and then incubated at 37 °C for 2 h.
2. Load the above reaction solution to a 100 kD cut-off centrifugal filter (Millipore, Bedford, MA) and centrifuge at 8000 g for 3 min (*see* **Note 3**).
3. Wash the centrifugal filter 4–6 times with 200 μL of water for each time.
4. Desorb the DNA retained on the filter by 100 μL of H_2O.

3.3 Enzymatic Digestion of DNA

1. Heat the Dpn I-treated genomic DNA (5 μg in 21 μL of H_2O) at 95 °C for 5 min and then chill on ice for 2 min for denaturation.
2. Add 3 μL of 10× enzymatic buffer, 1 μL of alkaline phosphatase, 2 μL of S1 nuclease, 2 μL of venom phosphodiesterase I, and 1 μL of DNase I. Incubate the mixture (30 μL) at 37 °C for 4 h (*see* **Note 4**).
3. Add 270 uL of H_2O and 300 μL of chloroform to the above reaction mixture. Vigorous vortex of the mixture for 3 min and then centrifugation at 12,000 g for 10 min.
4. Collect the aqueous layer and then add 300 μL of chloroform. Repeat the extraction for three times.
5. Collect the supernatant and lyophilize to dryness.
6. Redissolve the digested nucleosides in 100 μL of Milli-Q water for subsequent analysis. The sample can be analyzed immediately or stored at −20 °C until analysis.

3.4 Analysis of Nucleosides by LC-MS/MS

1. A Shimadzu LC-20 AD HPLC and an AB 3200 QTRAP mass spectrometer are used for analysis of nucleosides.
2. The HPLC separation of nucleosides is performed on a Hisep C18-T column. The flow rate is set at 0.2 mL/min and the

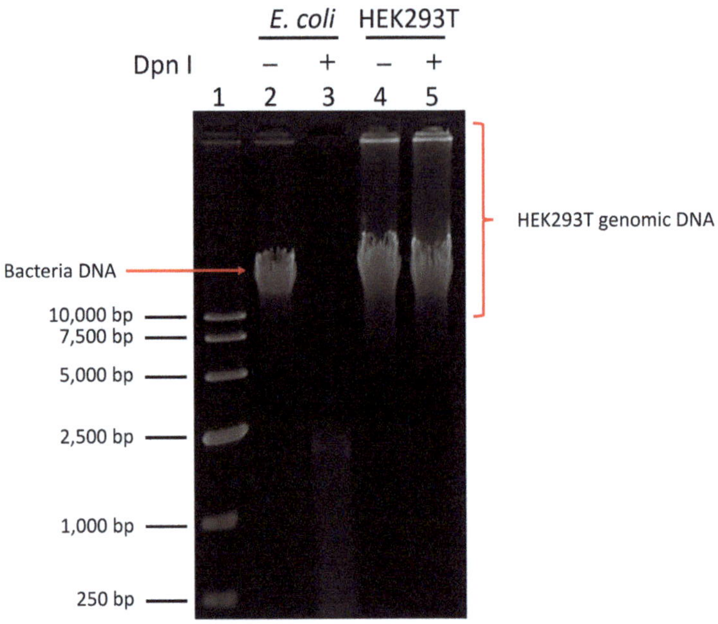

Fig. 2 Agarose gel electrophoresis analysis of DNA samples digested by Dpn I restriction enzyme. After Dpn I restriction enzyme cleavage, the DNA samples were loaded into a 0.8% agarose gel for electrophoresis with 0.5×TBE at 120 V for 1 h at room temperature. The gel was then staining by GelRed (Biotium, Inc., USA) and analyzed using the Pharos FX Molecular Imager (Bio-Rad, USA). Lane 1, DNA marker; Lane 2, bacterial DNA; Lane 3, bacterial DNA digested by Dpn I; Lane 4, HEK293T genomic DNA; Lane 5, HEK293T genomic DNA digested by Dpn I. (Reproduced from ref. [7] with permission from The Royal Society of Chemistry)

column temperature is 35 °C (*see* **Note 5**). Positive detection mode is used.

3. Use solvent A and solvent B as mobile phases with a gradient of 5 min 5% solvent B, 10 min 5–30% solvent B, 5 min 30–50% solvent B, 3 min 50–5% solvent B and 17 min 5% solvent B for the separation.
4. Monitor the nucleosides by multiple reaction monitoring (MRM) mode using the mass transitions (precursor ions→product ions) of T (243.1→127.1), dC (228.1→112.1), dA (252.1→136.1), dG (268.1→152.1), 6mA (266.1→150.2).
5. The MRM parameters of all nucleosides are optimized to achieve maximal detection sensitivity.
6. 6mA content in DNA is calculated using the following equation:

$$6mA\% = \frac{M_{6mA}}{M_{dA}} \times 100\%$$

where M_{6mA} is the molar quantity of 6mA, and M_{dA} is the molar quantity of dA determined in DNA sample.

7. The AB SCIEX Analyst 1.5 Software is used for data acquisition and processing.

3.5 Construction of Calibration Curve

1. To quantify 6mA in genomic DNA from mammals and plants, calibration curve is constructed by plotting the mean peak area ratio of 6mA/dA versus the mean molar ratio of 6mA/dA based on data obtained from triplicate measurements (Table 1).
2. The amounts of the analyte at signal/noise ratios of 3 and 10 are defined as limit of detection (LOD) and limit of quantification (LOQ) for 6mA, respectively (Table 1).

3.6 Determination of 6mA in Genomic DNA of Mammals and Plants

With the established method, we determine 6mA content in genomic DNA of mammals and plants. Figure 3 shows the typical extracted-ion chromatograms of 6mA from HEK293T cells, rat heart tissue, and Oryza sativa leaves. The results indicate that 6mA can be detected in all these samples. The measured content of 6mA in different samples is listed in Table 2. The method can efficiently remove the potential contamination from bacteria, which may benefit the study of the functions of 6mA in mammals and plants.

4 Notes

1. The solubility of dG or T is smaller than dA or dC. Some solid of dG or T standard may appear when they are stored at low temperature. The stock solution of dG or T should be heated to ensure completely dissolved before use.
2. The extracted genomic DNA may be contaminated by RNA if the OD260/280 value is greater than 1.90. Treating the isolated DNA with RNase can reduce the contamination of RNA.
3. The centrifugation time should be increased if there is liquid remained in the ultrafiltration tube after centrifugation.
4. If more than 10 μg of DNA sample is used, the amounts of enzymes and buffer should be increased accordingly.
5. Other types of C18 columns can be used for the separation of these nucleosides. To achieve the best resolution, the separation conditions should be optimized, including mobile phases, column temperature, and flow rate.

Table 1
Linearity, LOD and LOQ of 6mA by LC-MS/MS analysis. (Reproduced from ref. [7] with permission from The Royal Society of Chemistry)

	Regression line				
Linear range (6mA/dA, %)	Slope	Intercept	R^2	LOD (fmol)	LOQ (fmol)
0.00005–0.002	0.1506	0.000087	0.9983	0.42	1.38
0.002–0.5	0.1271	0.000134	0.9997		

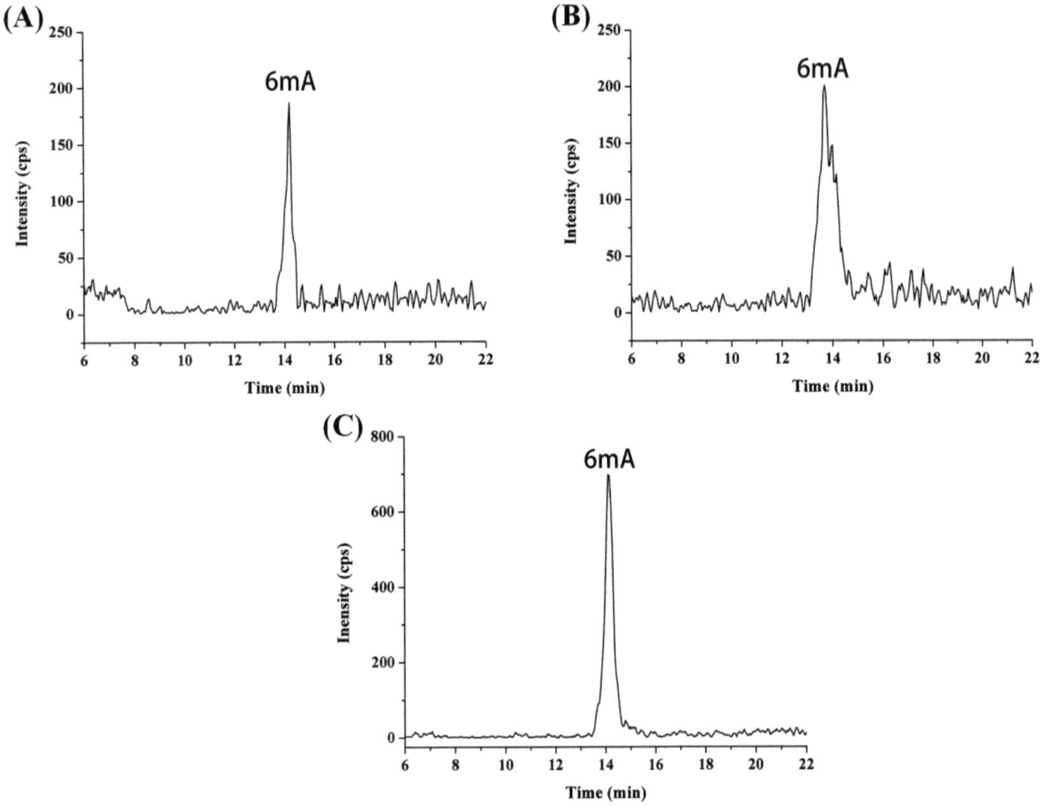

Fig. 3 Typical extracted-ion chromatograms for determination of 6mA in genomic DNA of mammals and plants. (**a**) HEK293T cells. (**b**) Rat heart tissue. (**c**) Oryza sativa leaves. (Reproduced from ref. [7] with permission from The Royal Society of Chemistry)

Table 2
Content of 6mA in genomic DNA of mammals and plants. (Reproduced from ref. [7] with permission from The Royal Society of Chemistry)

	Sample name	6mA/dA, %
Human cells	HEK293T	0.00017 ± 0.00001
	Jurkat-T	0.00023 ± 0.000001
Rat tissues	Heart	0.00032 ± 0.00003
	Liver	0.00013 ± 0.00001
	Spleen	0.00009 ± 0.00001
	Lung	0.00008 ± 0.000003
	Kidney	0.00006 ± 0.00001
	Brain	0.00013 ± 0.00002
	Subcutis	0.00013 ± 0.00002
Plants	Leaves of Oryza sativa	0.00077 ± 0.000004
	Kernels of Zea mays	0.00015 ± 0.00003

Acknowledgements

The work is supported by the National Natural Science Foundation of China (22074110, 21672166, 21721005).

References

1. Heyn H, Esteller M (2015) An adenine code for DNA: a second life for N6-methyladenine. Cell 161:710–713
2. Luo GZ, He C (2017) DNA N(6)-methyladenine in metazoans: functional epigenetic mark or bystander? Nat Struct Mol Biol 24:503–506
3. Fang G, Munera D, Friedman DI, Mandlik A, Chao MC, Banerjee O, Feng Z, Losic B, Mahajan MC, Jabado OJ, Deikus G, Clark TA, Luong K, Murray IA, Davis BM, Keren-Paz A, Chess A, Roberts RJ, Korlach J, Turner SW, Kumar V, Waldor MK, Schadt EE (2012) Genome-wide mapping of methylated adenine residues in pathogenic Escherichia coli using single-molecule real-time sequencing. Nat Biotechnol 30:1232–1239
4. Fu Y, Luo GZ, Chen K, Deng X, Yu M, Han D, Hao Z, Liu J, Lu X, Dore LC, Weng X, Ji Q, Mets L, He C (2015) N(6)-methyldeoxyadenosine marks active transcription start sites in chlamydomonas. Cell 161:879–892
5. Greer EL, Blanco MA, Gu L, Sendinc E, Liu J, Aristizabal-Corrales D, Hsu CH, Aravind L, He C, Shi Y (2015) DNA methylation on N (6)-adenine in C. elegans. Cell 161:868–878
6. Zhang G, Huang H, Liu D, Cheng Y, Liu X, Zhang W, Yin R, Zhang D, Zhang P, Liu J, Li C, Liu B, Luo Y, Zhu Y, Zhang N, He S, He C, Wang H, Chen D (2015) N(6)-methyladenine DNA modification in Drosophila. Cell 161:893–906
7. Huang W, Xiong J, Yang Y, Liu SM, Yuan BF, Feng YQ (2015) Determination of DNA adenine methylation in genomes of mammals and plants by liquid chromatography/mass spectrometry. RSC Adv 5:64046–64054
8. Xiao CL, Zhu S, He M, Chen ZQ, Chen Y, Yu G, Liu J, Xie SQ, Luo F, Liang Z, Wang DP, Bo XC, Gu XF, Wang K, Yan GR (2018) N(6)-methyladenine DNA modification in the human genome. Mol Cell 71:306–318.e307

9. Koziol MJ, Bradshaw CR, Allen GE, Costa AS, Frezza C, Gurdon JB (2016) Identification of methylated deoxyadenosines in vertebrates reveals diversity in DNA modifications. Nat Struct Mol Biol 23:24–30
10. Xie Q, Wu TP, Gimple RC, Li Z, Prager BC, Wu Q, Yu Y, Wang P, Wang Y, Gorkin DU, Zhang C, Dowiak AV, Lin K, Zeng C, Sui Y, Kim LJY, Miller TE, Jiang L, Lee CH, Huang Z, Fang X, Zhai K, Mack SC, Sander M, Bao S, Kerstetter-Fogle AE, Sloan AE, Xiao AZ, Rich JN (2018) N(6)-methyladenine DNA modification in glioblastoma. Cell 175:1228–1243
11. Wu TP, Wang T, Seetin MG, Lai Y, Zhu S, Lin K, Liu Y, Byrum SD, Mackintosh SG, Zhong M, Tackett A, Wang G, Hon LS, Fang G, Swenberg JA, Xiao AZ (2016) DNA methylation on N(6)-adenine in mammalian embryonic stem cells. Nature 532:329–333
12. Liang Z, Shen L, Cui X, Bao S, Geng Y, Yu G, Liang F, Xie S, Lu T, Gu X, Yu H (2018) DNA N(6)-adenine methylation in Arabidopsis thaliana. Dev Cell 45:406–416.e403
13. Zhou C, Wang C, Liu H, Zhou Q, Liu Q, Guo Y, Peng T, Song J, Zhang J, Chen L, Zhao Y, Zeng Z, Zhou DX (2018) Identification and analysis of adenine N(6)-methylation sites in the rice genome. Nat Plants 4:554–563
14. Liu J, Zhu Y, Luo GZ, Wang X, Yue Y, Wang X, Zong X, Chen K, Yin H, Fu Y, Han D, Wang Y, Chen D, He C (2016) Abundant DNA 6mA methylation during early embryogenesis of zebrafish and pig. Nat Commun 7:13052
15. Luo GZ, Blanco MA, Greer EL, He C, Shi Y (2015) DNA N(6)-methyladenine: a new epigenetic mark in eukaryotes? Nat Rev Mol Cell Biol 16:705–710
16. Schiffers S, Ebert C, Rahimoff R, Kosmatchev O, Steinbacher J, Bohne AV, Spada F, Michalakis S, Nickelsen J, Muller M, Carell T (2017) Quantitative LC-MS provides no evidence for m(6) dA or m(4) dC in the genome of mouse embryonic stem cells and tissues. Angew Chem Int Ed Engl 56:11268–11271
17. Douvlataniotis K, Bensberg M, Lentini A, Gylemo B, Nestor CE (2020) No evidence for DNA N (6)-methyladenine in mammals. Sci Adv 6:eaay3335
18. Musheev MU, Baumgartner A, Krebs L, Niehrs C (2020) The origin of genomic N(6)-methyl-deoxyadenosine in mammalian cells. Nat Chem Biol 16:630–634
19. Hao Z, Wu T, Cui X, Zhu P, Tan C, Dou X, Hsu KW, Lin YT, Peng PH, Zhang LS, Gao Y, Hu L, Sun HL, Zhu A, Liu J, Wu KJ, He C (2020) N(6)-deoxyadenosine methylation in mammalian mitochondrial DNA. Mol Cell 78:382–395.e388
20. Chen B, Yuan BF, Feng YQ (2019) Analytical methods for deciphering RNA modifications. Anal Chem 91:743–756
21. Feng Y, Xie NB, Tao WB, Ding JH, You XJ, Ma CJ, Zhang X, Yi C, Zhou X, Yuan BF, Feng YQ (2020) Transformation of 5-carboxylcytosine to cytosine through C–C bond cleavage in human cells constitutes a novel pathway for DNA demethylation. CCS Chem 2:994–1008
22. Xiong J, Yuan BF, Feng YQ (2019) Mass spectrometry for investigating the effects of toxic metals on nucleic acid modifications. Chem Res Toxicol 32:808–819
23. Yuan BF (2020) Assessment of DNA epigenetic modifications. Chem Res Toxicol 33:695–708
24. Cheng QY, Xiong J, Ma CJ, Dai Y, Ding JH, Liu FL, Yuan BF, Feng YQ (2020) Chemical tagging for sensitive determination of uridine modifications in RNA. Chem Sci 11:1878–1891
25. Qi CB, Jiang HP, Xiong J, Yuan BF, Feng YQ (2019) On-line trapping/capillary hydrophilic-interaction liquid chromatography/mass spectrometry for sensitive determination of RNA modifications from human blood. Chin Chem Lett 30:553–557
26. Lan MD, Yuan BF, Feng YQ (2019) Deciphering nucleic acid modifications by chemical derivatization-mass spectrometry analysis. Chin Chem Lett 30:1–6
27. Chen ML, Shen F, Huang W, Qi JH, Wang Y, Feng YQ, Liu SM, Yuan BF (2013) Quantification of 5-methylcytosine and 5-hydroxymethylcytosine in genomic DNA from hepatocellular carcinoma tissues by capillary hydrophilic-interaction liquid chromatography/quadrupole TOF mass spectrometry. Clin Chem 59:824–832
28. Xiong J, Ye TT, Ma CJ, Cheng QY, Yuan BF, Feng YQ (2019) N6-hydroxymethyladenine: a hydroxylation derivative of N6-methyladenine in genomic DNA of mammals. Nucleic Acids Res 47:1268–1277
29. Tang Y, Chu JM, Huang W, Xiong J, Xing XW, Zhou X, Feng YQ, Yuan BF (2013) Hydrophilic material for the selective enrichment of 5-hydroxymethylcytosine and its liquid chromatography-tandem mass spectrometry detection. Anal Chem 85:6129–6135
30. Xing XW, Liu YL, Vargas M, Wang Y, Feng YQ, Zhou X, Yuan BF (2013) Mutagenic and cytotoxic properties of oxidation products of 5-methylcytosine revealed by next-generation sequencing. PLoS One 8:e72993

31. Tang Y, Xiong J, Jiang HP, Zheng SJ, Feng YQ, Yuan BF (2014) Determination of oxidation products of 5-methylcytosine in plants by chemical derivatization coupled with liquid chromatography/tandem mass spectrometry analysis. Anal Chem 86:7764–7772

Part III

Detection of 5hmU and 5fU

Chapter 12

Isotope-Dilution Liquid Chromatography–Tandem Mass Spectrometry for Detection of 5-Hydroxymethyluracil and 5-Formyluracil in DNA

Yu Liu and Jin Wang

Abstract

5-Hydroxymethyluracil (5hmU) and 5-formyluracil (5fU) are mutagenic and major oxidized nucleobases in DNA that are derived from thymine (T) or 5-methylcytosine (5mC). 5hmU, and possibly 5fU as well, in DNA may also assume an epigenetic role. The analysis of 5hmU and 5fU in DNA has suffered from multiple drawbacks such as a lack of specificity, artifacts generated in sample preparation, as well as poor applicability to analysis of nucleoside forms of DNA base damage. To overcome these problems, we developed a capillary liquid chromatography–electrospray ionization–tandem mass spectrometry (LC-ESI-MS^3) method that is combined with the stable isotope-dilution technique, which allows quantification of 5hmU, 5fU, and several other oxidation products in DNA with high sensitivity, specificity, and accuracy. In this chapter, we describe in detail the protocols from DNA extraction, through DNA digestion and off-line HPLC enrichment, to LC-MS^3 analysis. The method is readily applicable to other modified nucleosides in DNA.

Keywords DNA damage, Liquid chromatography, Mass spectrometry, Isotope dilution, 5-Hydroxymethyluracil, 5-Formyluracil, Oxidative stress

1 Introduction

Reactive oxygen species (ROS) can be induced endogenously or by exogenous sources, and excess generation of ROS in vivo can cause damage to DNA and other biological molecules [1–3]. 5-hydroxymethyluracil (5hmU) and 5-formyluracil (5fU) are two major DNA lesions generated by ROS [4–7].

5hmU resulted from T oxidation pairs with adenine (A), while 5hmU generated by deamination and oxidation of 5mC pairs with guanine (G). In earlier studies of 5hmU, focus was placed on its role as a DNA lesion formed by oxidation of T and revealed that 5hmU is weakly mutagenic [8], is not a block to DNA polymerases (Pols) [9], and pairs with A in Watson–Crick geometry [10]. Human cell extracts showed a much lower excision activity toward 5hmU in 5hmU:A pair than that in 5-hmU:G pair

Bi-Feng Yuan (ed.), *DNA Modification Detection Methods*, Springer Protocols Handbooks,
https://doi.org/10.1007/978-1-0716-1229-3_12,

[11]. 5hmU:G is excisable by glycosylases such as SMUG1, NEIL1, TDG, and MBD4 [12–15]. SMUG1 also showed 5-hmU:A excision activity, albeit with lower activity [15]. These attributes of 5hmU are consistent with the recently hypothesized role of this modified nucleoside as an essential epigenetic mediator [16–19].

5fU is not a strong block to DNA Pols [20]. Although any of the four dNTPs can be incorporated opposite this lesion by high-fidelity DNA Pols in vitro, 5fU is weakly mutagenic, inducing T → A and T → G transversions in simian COS-7 cells [21]. The repair of 5fU is predominantly performed via the BER pathway. In *Escherichia coli* and mammalian cells, 5fU was primarily removed from DNA by AlkA [22–24] and SMUG1 [25–27], respectively. Similar to 5hmU, 5fU in DNA exists as 5fU:A or 5fU:G, with the former form repaired by AlkA [23, 24], and the latter removed by AlkA and the MMR pathway [22, 28]. 5-formyl-2′-deoxycytidine (5fdC) was found to undergo spontaneous deamination, at rates higher than 5-hydroxymethyl-2′-deoxycytidine (5hmdC), 2′-deoxycytidine (dC), and 5-methyl-2′-deoxycytidine (5mdC), to form 5-formyl-2′-deoxyuridine (5fdU) at physiological pH [29], suggesting that 5fU in DNA in vivo may in part arise from spontaneous deamination of 5fC. This implies that 5fU may assume an epigenetic role like 5hmU and 5fC.

Analysis of low-level DNA damage products including 5hmU and 5fU, in biological samples, is a challenging task that requires high sensitivity and specificity to achieve unequivocal identification and quantification. Several analytical methods to quantify 5hmU and 5fU in DNA have been developed. Earlier studies employed high-performance liquid chromatography (HPLC) in combination with ^{32}P-postlabeling [30] or ^{3}H-labeling [24, 31–33], which allows the measurement of various DNA damage products including 5hmU and 5fU with high specificity and sensitivity. While sensitive, the radiolabeling approach generally lacks chemical specificity and internal standards that serve as a control for loss and artifacts during sample preparation.

Alternatively, mass spectrometry (MS)-based methods that employ isotopically labeled internal standards provide unequivocal identification and quantification [34]. Gas chromatography coupled with mass spectrometry (GC-MS) provides high sensitive quantification of many types of DNA lesions including 5hmU and 5fU [35, 36], though this approach is not as readily applicable to nucleoside forms of DNA base damage as LC-MS and suffers from overestimation of oxidized DNA nucleosides due to artifactual oxidation [37, 38]. More recently, reversed-phase HPLC coupled with isotope–dilution electrospray ionization–tandem mass spectrometry (LC-ESI-MS/MS) has become popular and the method of choice for the analysis of 5hmU, 5fU, and other oxidized nucleosides in DNA [39–42]. This method provides excellent separation of nucleosides, readily released from DNA through enzymatic digestion, and highly sensitive, specific, and accurate quantification of nucleosides. The methodology reported by Wang et al. [42], which involves the off-line HPLC enrichment (OHE) of 5hmU, 5fU, and several others modified nucleosides prior to capillary LC-MS/MS/MS (LC-MS^3) analysis, provides the basis for the procedure described here. OHE is considered as one of the best methods for isolating trace amounts of damaged nucleosides from DNA digestion mixtures prior to MS analysis as it does not use solid-phase particles which can cause clogging of nano-HPLC columns [43], removes most of the highly abundant unmodified nucleosides as well as buffer salts in the enzymatic digest, and affords nearly quantitative analyte recovery. Additionally, capillary flow-LC-MS^3 analysis provides increased specificity and sensitivity compared to microflow-LC-MS/MS analysis. The procedure provides quantification of 5hmU, 5fU, and several other oxidized nucleosides in rat liver and brain DNA with high specificity, sensitivity, and accuracy. The method is readily applicable to other tissue/cell types and other modified nucleosides in DNA.

2 Materials

All reagents and enzymes used are available from Sigma–Aldrich if not otherwise stated.

2.1 Enzymes

1. 1 unit/μL nuclease P1: dissolve nuclease P1 with Milli-Q grade water.
2. 0.01 unit/μL phosphodiesterase 1: dilute phosphodiesterase 1 with Milli-Q grade water.
3. 0.005 unit/μL phosphodiesterase 2: dilute phosphodiesterase 2 with Milli-Q grade water.
4. 1 unit/μL alkaline phosphatase calf intestinal (CIP): dilute CIP with Milli-Q grade water.
5. 20 mg/mL proteinase K.
6. 10 mg/mL RNase A.
7. 25 units/μL RNase T1.

2.2 Buffers and Other Reagents

1. DNA digestion buffer 1: 300 mM sodium acetate (pH 5.6), 10 mM zinc chloride.
2. DNA digestion buffer 2: 500 mM Tris–HCl, pH 8.9.
3. Lysis buffer: 20 mM Tris (pH 8.1), 400 mM NaCl, 20 mM EDTA, 1% SDS.
4. Chloroform/isoamyl alcohol (24:1, v/v).
5. Ethanol.

6. Saturated NaCl solution.
7. Tris-EDTA (TE) buffer, pH 8.0.
8. 10 mM erythro-9-(2-hydroxy-3-nonyl)adenine (EHNA) hydroxide: dissolve with Milli-Q grade water.
9. 10 M ammonium acetate: dissolve with Milli-Q grade water.
10. 1 M formic acid: dilute with Milli-Q grade water.
11. Chloroform.
12. 70% ethanol.

2.3 Liquid Chromatography

1. 10 mM ammonium formate (solution A): dissolve with Milli-Q grade water.
2. 0.1% (v/v) formic acid in H_2O (solution C): dilute with Milli-Q grade water.
3. 0.1% (v/v) formic acid in methanol (solution D): dilute with methanol.

2.4 Nucleoside Standards

All nucleosides standards are dissolved at appropriate concentrations in Milli-Q grade water.

1. 5-hydroxymethyl-2′-deoxyuridine (5hmdU).
2. 5-formyl-2′-deoxyuridine (5fdU).
3. [1,3-^{15} N_2–2′-D]-5-hydroxymethyl-2′-deoxyuridine ($^{15}N_2$, D-5hmdU) [44].
4. [1,3-^{15} N_2–2′-D]-5-formyl-2′-deoxyuridine ($^{15}N_2$, D-5fdU) [44].
5. 2′-deoxyadenosine (dA).
6. 2′-deoxyguanosine (dG).
7. 2′-deoxycytidine (dC).
8. Thymidine (dT).
9. Adenosine (A).
10. Guanosine (G).
11. Cytidine (C).
12. Uridine (U).

2.5 Equipment

1. A Beckman System Gold HPLC system equipped with a quaternary pumping system and a UV–VIS detector.
2. Reversed-phase column Alltima HP C18 column (4.6 × 250 mm, 5 μm, Grace Davison, Deerfield, IL).
3. Reversed-phase Zorbax SB-C18 column (0.5 × 250 mm, 5 μm, Agilent).
4. A NanoDrop 1000 spectrophotometer system (Thermo Fisher).

5. An Agilent 1100 capillary HPLC system equipped with a quaternary pumping system.
6. An LTQ linear ion-trap mass spectrometer (Thermo Fisher Scientific).
7. A Savant SpeedVac Plus SC110A system (Thermo) equipped with a cold trap.

3 Methods

The entire procedure from DNA extraction to LC-MS^3 analysis is summarized in Fig. 1. This procedure takes approximately 4 days of elapsed time and consists of the following major steps:

1. Extraction of genomic DNA (approximately 20 h of elapsed time).
2. Enzymatic digestion of genomic DNA (2 days of elapsed time).
3. Off-line HPLC enrichment (2 h elapsed time/sample).
4. Concentration of HPLC fractions with SpeedVac (half day of elapsed time).
5. LC-MS^3 analysis (1 h of elapsed time/sample).

3.1 Extraction of Genomic DNA

1. Grind the liver or brain tissue (approximately 200 mg) from a Long–Evans Cinnamon (LEC) or Long–Evan Agouti (LEA) rat under liquid N_2 into fine powders in a mortar.
2. Transfer the powders immediately into a 50 mL polypropylene centrifuge tube, and then add 5 mL of the lysis buffer and 25 μL of proteinase K (20 mg/mL).
3. Incubate the tube in a water bath at 55 °C for 16 h.

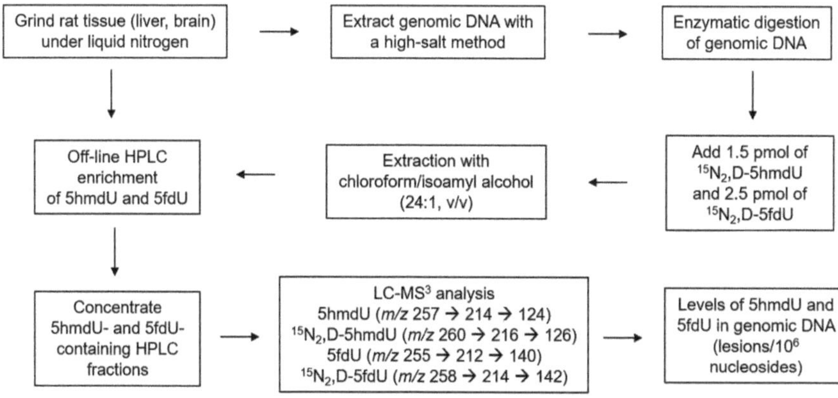

Fig. 1 Summary of the procedure from DNA extraction to LC-MS^3 analysis

4. Add 2.5 mL of saturated NaCl solution into the mixture, vortex for 1 min, and then incubate at 55 °C for 15 min.
5. Centrifuge the mixture at approximately 10,000 rpm for 30 min, collect the supernatant, and centrifuge again under the same conditions.
6. Add 2 volumes of cold ethanol into the resulting supernatant to precipitate the nucleic acids.
7. Transfer the precipitated nucleic acid pellet into a new 1.5 mL polypropylene microcentrifuge tube, which is then washed with cold 70% ethanol twice and reconstituted in 100 μL of Milli-Q grade water.
8. Add 2.5 μL of RNase A (10 mg/mL) and 1 μL of RNase T1 (25 units/μL) to the nucleic acid mixture, incubate at 37 °C for 1 h, and then extract with an equal volume of chloroform/isoamyl alcohol (24:1, v/v) twice.
9. Precipitate the DNA from the upper layer by adding 1/10 volume of 10 M ammonium acetate and 2 volume of cold ethanol. Store the mixture at −20 °C overnight.
10. Centrifuge the mixture at 5000 rpm at 4 °C for 10 min and transfer the precipitated nucleic acid pellet into a new 1.5 mL polypropylene microcentrifuge tube, which is then washed with cold 70% ethanol twice.
11. Centrifuge the mixture at 7000 rpm at 4 °C for 15 min.
12. Wash the resulting DNA pellet twice with 70% cold ethanol and allow it to air-dry at room temperature overnight.
13. Dissolve the DNA pellet in 50 μL of TE buffer (*see* **Note 1**), with the concentration of the genomic DNA quantified on a NanoDrop 1000 spectrophotometer system. The DNA solution is then used immediately or stored at −20 °C until use.

3.2 Enzymatic Digestion of Genomic DNA

1. 20 nmol of EHNA (an adenine deaminase), 8 units of nuclease P1, 0.01 unit of phosphodiesterase 2, and a 20 μL solution containing DNA Digestion Buffer 1 (*see* **Note 1**) is added to 80 μg of DNA and vortex briefly to mix well.
2. Incubate the mixture at 37 °C for 48 h.
3. Add 40 μL of DNA digestion buffer 2, 0.02 unit of phosphodiesterase 1, and 8 units of CIP to the digestion mixture and vortex briefly to mix well.
4. Incubate the mixture at 37 °C for 2 h, which is subsequently neutralized by addition of 1 μL of 1 M formic acid (*see* **Note 2**).
5. Add the stable-isotope labeled internal standards to the mixture, which include 1.5 pmol $^{15}N_2$, D-5hmdU, and 2.5 pmol $^{15}N_2$, D-5fdU, and vortex briefly to mix well.

6. Remove the enzymes in the mixture subsequently by extraction with approximately 90 μL of chloroform.
7. Isolate and SpeedVac the resulting aqueous layer to dryness, which is subsequently reconstituted in 100 μL doubly distilled water (*see* **Note 3**).

3.3 Off-line HPLC Enrichment

The reconstituted DNA digest is then subjected to off-line HPLC separation for the enrichment of 5hmdU and 5fdU.

1. Precondition a 4.6 mm × 250 mm Alltima HP C18 column (particle size, 5 μm) by allowing solution A to pass through it for approximately 90 min at a flow rate of 1 mL/min.
2. Inject 100 μL of the reconstituted DNA digest onto the column via a manual injection valve.
3. The column is eluted with a gradient of 42 min 100% solution A, 1 min 0–5% methanol (solution B), 32 min 5% B, 5 min 5–20% B, and 10 min 20% B is employed for each HPLC run. A typical HPLC trace is shown in Fig. 2. A minor RNA contamination (<5% for most DNA samples) is often observed for the DNA samples according to the HPLC traces. Correction for RNA contamination is made for each of the DNA samples analyzed according to the peak areas of A, G, dA, and dG in the HPLC traces.
4. The HPLC fractions eluting at 17.0–20.0 min and 32.5–36.0 min are pooled for 5hmdU and 5fdU, respectively.
5. SpeedVac the collected fractions to dryness and reconstitute in 12 μL of Milli-Q grade H_2O.

3.4 LC-MS^3 Analysis

The reconstituted HPLC fractions are subsequently injected into a capillary LC-LTQ linear ion-trap mass spectrometer system for LC-MS^3 analysis.

1. The mass spectrometer is tuned in negative mode by using an ESI negative ion mode calibration solution.
2. Precondition a 0.5 mm × 250 mm Zorbax SB-C18 column (particle size, 5 μm) by allowing solution C to pass through it at a flow rate of 8 μL/min for 30–40 min.
3. Inject the reconstituted fractions onto the column by using a 10 μL injection syringe (*see* **Note 4**).
4. The column is eluted with a gradient of 5 min 0–20% solution D, 21 min 20–35% D, and 9 min 35–80% D.
5. MS^3 analysis (*see* **Note 5**) is performed in negative ion modes for the following transitions of the $[M-H]^-$ ions of unlabeled and isotope-labeled 5hmdU and 5fdU: 5hmdU (*m/z* 257 → 214 → 124), $^{15}N_2$, D-5hmdU (*m/z* 260 → 216 → 126), 5fdU (*m/z* 255 → 212 → 140), and

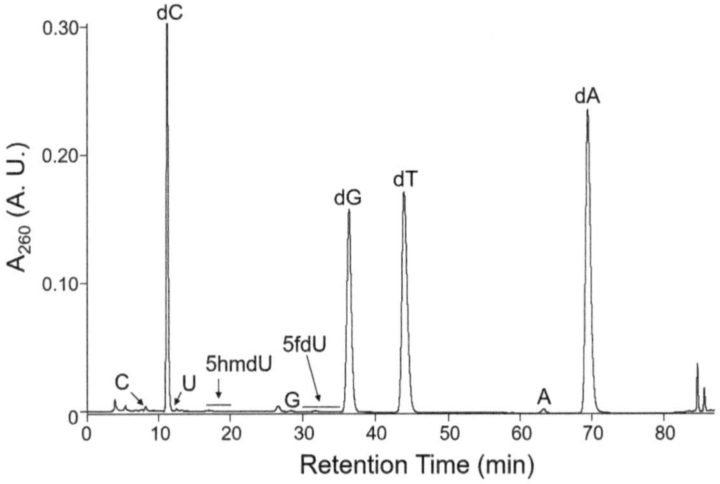

Fig. 2 Representative trace for off-line HPLC enrichment of 5hmdU and 5fdU from the digestion mixture of genomic DNA from rat tissue. (Reproduced from [42] with permission from American Chemical Society)

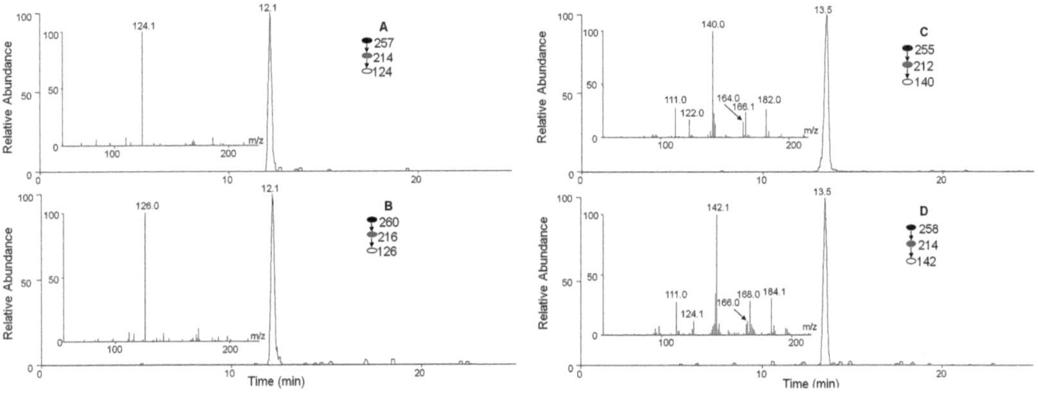

Fig. 3 Representative selected-ion chromatograms (SICs) for monitoring the *m/z* 257 → 214 → 124 (**a**) for unlabeled 5hmdU, *m/z* 260 → 216 → 126 (**b**) for $^{15}N_2$, D-5hmdU, *m/z* 255 → 212 → 140 (**c**) for unlabeled 5fdU, and *m/z* 258 → 214 → 142 (**d**) for $^{15}N_2$, D-5fdU of the digestion mixture of genomic DNA from rat tissue

$^{15}N_2$, D-5fdU (*m/z* 258 → 214 → 142). Typical selected-ion chromatograms (SICs) are depicted in Fig. 3.

6. The levels of 5hmdU and 5fdU in each tissue sample are calculated from the amounts (in moles) of lesions obtained from the LC-MS^3 analyses (according to the peak areas in the corresponding SICs for the analytes and internal standards, the amounts of internal standards added, and the calibration curves) and the amount of canonical nucleosides (in moles, according to UV absorbance at 260 nm) in the genomic DNA used for the digestion.

4 Notes

1. The volume of this buffer can be variable.
2. This neutralization step is employed to avoid deviation of elution pH and shortening of HPLC column lifetime due to the injection of alkaline DNA digests during the off-line HPLC enrichment.
3. A slightly smaller volume of doubly distilled water can also be used. Water of higher purity should also be applicable, such as Milli-Q grade water.
4. Typically 8 μL of each of the reconstituted fractions is injected for analysis, with the remainder reserved for re-analysis if necessary.
5. Prior to analysis of 5hmdU and 5fdU by LC-MS^3, two working parameters of the LTQ mass spectrometer, i.e., activation Q and normalized collision energy, were optimized for detection of these two lesions with optimal sensitivity.

Acknowledgments

The development of the methodology used on which this procedure was prepared was supported by the National Institutes of Health (Grants R01 CA101864, R01 DK082779, R01 DK071111, and P30-DK41296). The preparation of this procedure was supported by Inner Mongolia University with a grant from its "Outstanding Young Talents" Program (to YL) and a grant from its "Steed Plan" High-Level Talents Program (to JW).

References

1. Marnett LJ (2000) Oxyradicals and DNA damage. Carcinogenesis 21:361–370
2. Lindahl T (1999) DNA lesions generated in vivo by reactive oxygen species, their accumulation and repair. NATO ASI Ser Ser A 302:251–257
3. Finkel T, Holbrook NJ (2000) Oxidants, oxidative stress and the biology of ageing. Nature 408:239–247
4. Cadet J, Bellon S, Douki T et al (2004) Radiation-induced DNA damage: formation, measurement, and biochemical features. J Environ Pathol Toxicol Oncol 23:33–43
5. Kasai H, Iida A, Yamaizumi Z et al (1990) 5-Formyldeoxyuridine: a new type of DNA damage induced by ionizing radiation and its mutagenicity to salmonella strain TA102. Mutat Res 243:249–253
6. Olinski R, Starczak M, Gackowski D (2016) Enigmatic 5-hydroxymethyluracil: oxidatively modified base, epigenetic mark or both? Mutat Res Rev Mutat Res 767:59–66
7. Teebor GW, Boorstein RJ, Cadet J (1988) The repairability of oxidative free radical mediated damage to DNA: a review. Int J Radiat Biol 54:131–150
8. Levy DD, Teebor GW (1991) Site directed substitution of 5-hydroxymethyluracil for thymine in replicating phi X-174am3 DNA via synthesis of 5-hydroxymethyl-2-′-deoxyuridine-5′-triphosphate. Nucleic Acids Res 19:3337–3343
9. Herrala AM, Vilpo JA (1989) Template-primer activity of 5-(hydroxymethyl)uracil-containing DNA for prokaryotic and eukaryotic DNA and

RNA polymerases. Biochemistry 28:8274–8277
10. Mellac S, Fazakerley GV, Sowers LC (1993) Structures of base pairs with 5-(hydroxymethyl)-2′-deoxyuridine in DNA determined by NMR spectroscopy. Biochemistry 32:7779–7786
11. Rusmintratip V, Sowers LC (2000) An unexpectedly high excision capacity for mispaired 5-hydroxymethyluracil in human cell extracts. Proc Natl Acad Sci U S A 97:14183–14187
12. Hashimoto H, Hong S, Bhagwat AS et al (2012) Excision of 5-hydroxymethyluracil and 5-carboxylcytosine by the thymine DNA glycosylase domain: its structural basis and implications for active DNA demethylation. Nucleic Acids Res 40:10203–10214
13. Morera S, Grin I, Vigouroux A et al (2012) Biochemical and structural characterization of the glycosylase domain of MBD4 bound to thymine and 5-hydroxymethyuracil-containing DNA. Nucleic Acids Res 40:9917–9926
14. Cortellino S, Xu J, Sannai M et al (2011) Thymine DNA glycosylase is essential for active DNA demethylation by linked deamination-base excision repair. Cell 146:67–79
15. Wibley JE, Waters TR, Haushalter K et al (2003) Structure and specificity of the vertebrate anti-mutator uracil-DNA glycosylase SMUG1. Mol Cell 11:1647–1659
16. Modrzejewska M, Gawronski M, Skonieczna M et al (2016) Vitamin C enhances substantially formation of 5-hydroxymethyluracil in cellular DNA. Free Radic Biol Med 101:378–383
17. Kawasaki F, Martinez Cuesta S, Beraldi D et al (2018) Sequencing 5-hydroxymethyluracil at single-base resolution. Angew Chem Int Ed Engl 57:9694–9696
18. Kawasaki F, Beraldi D, Hardisty RE et al (2017) Genome-wide mapping of 5-hydroxymethyluracil in the eukaryote parasite Leishmania. Genome Biol 18:23
19. Pfaffeneder T, Spada F, Wagner M et al (2014) Tet oxidizes thymine to 5-hydroxymethyluracil in mouse embryonic stem cell DNA. Nat Chem Biol 10:574–581
20. Klungland A, Paulsen R, Rolseth V et al (2001) 5-Formyluracil and its nucleoside derivatives confer toxicity and mutagenicity to mammalian cells by interfering with normal RNA and DNA metabolism. Toxicol Lett 119:71–78
21. Kamiya H, Murata-Kamiya N, Karino N et al (2002) Induction of T --> G and T --> A transversions by 5-formyluracil in mammalian cells. Mutat Res 513:213–222
22. Terato H, Masaoka A, Kobayashi M et al (1999) Enzymatic repair of 5-formyluracil. II. Mismatch formation between 5-formyluracil and guanine during dna replication and its recognition by two proteins involved in base excision repair (AlkA) and mismatch repair (MutS). J Biol Chem 274:25144–25150
23. Masaoka A, Terato H, Kobayashi M et al (1999) Enzymatic repair of 5-formyluracil. I. Excision of 5-formyluracil site-specifically incorporated into oligonucleotide substrates by alka protein (*Escherichia coli* 3-methyladenine DNA glycosylase II). J Biol Chem 274:25136–25143
24. Bjelland S, Birkeland NK, Benneche T et al (1994) DNA glycosylase activities for thymine residues oxidized in the methyl group are functions of the AlkA enzyme in *Escherichia coli*. J Biol Chem 269:30489–30495
25. Knaevelsrud I, Slupphaug G, Leiros I et al (2009) Opposite-base dependent excision of 5-formyluracil from DNA by hSMUG1. Int J Radiat Biol 85:413–420
26. Matsubara M, Masaoka A, Tanaka T et al (2003) Mammalian 5-formyluracil-DNA glycosylase. 1. Identification and characterization of a novel activity that releases 5-formyluracil from DNA. Biochemistry 42:4993–5002
27. Masaoka A, Matsubara M, Hasegawa R et al (2003) Mammalian 5-formyluracil-DNA glycosylase. 2. Role of SMUG1 uracil-DNA glycosylase in repair of 5-formyluracil and other oxidized and deaminated base lesions. Biochemistry 42:5003–5012
28. Liu P, Burdzy A, Sowers LC (2003) Repair of the mutagenic DNA oxidation product, 5-formyluracil. DNA Repair 2:199–210
29. Schiesser S, Pfaffeneder T, Sadeghian K et al (2013) Deamination, oxidation, and C-C bond cleavage reactivity of 5-hydroxymethylcytosine, 5-formylcytosine, and 5-carboxycytosine. J Am Chem Soc 135:14593–14599
30. Cadet J, Odin F, Mouret JF et al (1992) Chemical and biochemical postlabeling methods for singling out specific oxidative DNA lesions. Mutat Res 275:343–354
31. Teebor GW, Frenkel K, Goldstein MS (1984) Ionizing radiation and tritium transmutation both cause formation of 5-hydroxymethyl-2-′-deoxyuridine in cellular DNA. Proc Natl Acad Sci U S A 81:318–321
32. Frenkel K, Cummings A, Solomon J et al (1985) Quantitative determination of the 5-(hydroxymethyl)uracil moiety in the DNA of gamma-irradiated cells. Biochemistry 24:4527–4533

33. Bjelland S, Eide L, Time RW et al (1995) Oxidation of thymine to 5-formyluracil in DNA: mechanisms of formation, structural implications, and base excision by human cell free extracts. Biochemistry 34:14758–14764
34. Dizdaroglu M, Jaruga P, Birincioglu M et al (2002) Free radical-induced damage to DNA: mechanisms and measurement. Free Radic Biol Med 32:1102–1115
35. Douki T, Delatour T, Paganon F et al (1996) Measurement of oxidative damage at pyrimidine bases in gamma-irradiated DNA. Chem Res Toxicol 9:1145–1151
36. Mori T, Dizdaroglu M (1994) Ionizing radiation causes greater DNA base damage in radiation-sensitive mutant M10 cells than in parent mouse lymphoma L5178Y cells. Radiat Res 140:85–90
37. Dizdaroglu M (1998) Facts about the artifacts in the measurement of oxidative DNA base damage by gas chromatography-mass spectrometry. Free Radic Res 29:551–563
38. Cadet J, Douki T, Ravanat JL (1997) Artifacts associated with the measurement of oxidized DNA bases. Environ Health Perspect 105:1034–1039
39. Cadet J, Douki T, Frelon S et al (2002) Assessment of oxidative base damage to isolated and cellular DNA by HPLC-MS/MS measurement. Free Radic Biol Med 33:441–449
40. Frelon S, Douki T, Ravanat JL et al (2000) High-performance liquid chromatography--tandem mass spectrometry measurement of radiation-induced base damage to isolated and cellular DNA. Chem Res Toxicol 13:1002–1010
41. Hua Y, Wainhaus SB, Yang Y et al (2001) Comparison of negative and positive ion electrospray tandem mass spectrometry for the liquid chromatography tandem mass spectrometry analysis of oxidized deoxynucleosides. J Am Soc Mass Spectrom 12:80–87
42. Wang J, Yuan B, Guerrero C et al (2011) Quantification of oxidative DNA lesions in tissues of Long-Evans Cinnamon rats by capillary high-performance liquid chromatography-tandem mass spectrometry coupled with stable isotope-dilution method. Anal Chem 83:2201–2209
43. Tretyakova N, Goggin M, Sangaraju D et al (2012) Quantitation of DNA adducts by stable isotope dilution mass spectrometry. Chem Res Toxicol 25:2007–2035
44. Hong H, Cao H, Wang Y et al (2006) Identification and quantification of a guanine-thymine intrastrand cross-link lesion induced by Cu (II)/H_2O_2/ascorbate. Chem Res Toxicol 19:614–621

Chapter 13

Detection of 5-Formylcytosine and 5-Formyluracil Based on Photo-Assisted Domino Reaction

Qian Zhou, Kun Li, Kang-Kang Yu, and Xiao-Qi Yu

Abstract

5-Formylcytosine (5fC) is a recently identified DNA modification in mammalian genome, derived from 5-methylcytosine (5mC) by TET-mediated oxidation. It has been demonstrated to be a critical intermediate during active demethylation and may also be a stable epigenetic marker playing independent biological roles. To further understanding its complicated epigenetic functions, quantitative analysis methods with high selectivity and sensitivity for 5fC are urgently required. In this respect, how to distinguish 5fC from 5-formyluracil (5fU) to achieve higher accuracy remains challenging because the latter one is more reactive. We take advantage of the slight difference in structure between 5fC and 5fU and introduce a Wittig-initiated photocatalytic triple domino reaction to selectively switch on 5fC using a commercially available phosphorus ylide. Such a fluorescence-based detection approach is easy to operate and short time-consuming, but highly selective, enabling accurately qualitative and quantitative analysis of 5fC.

Keywords 5-Formylcytosine, 5-Formyluracil, Wittig reaction, Photocyclization, Domino reaction, Fluorescence

1 Introduction

Cytosine methylation at CpG islands by DNA methyltransferases using S-adenosylmethionine (SAM) as a source of the methyl group produces 5-methylcytosine (5mC), which is the most common epigenetic modification in mammalian genomes [1]. Ten-eleven translocation (Tet) family of dioxygenases can further oxidize 5mC to 5-hydroxymethylcytosine (5hmC), 5-formylcytosine (5fC), and 5-carboxylcytosine (5caC) in a stepwise manner. Recognition and cleavage of 5fC and 5caC with thymine-DNA glycosylase (TDG) followed by base-excision repair (BER) results in restoration of canonical cytosine [2], leading to an active DNA demethylation pathway [3]. Recent studies revealed that the majority of 5fC are stable or at least semi-permanent, in mammalian cells, suggesting additional roles of 5fC in DNA as a potential epigenetic mark that go beyond being a demethylation intermediate

Bi-Feng Yuan (ed.), *DNA Modification Detection Methods*, Springer Protocols Handbooks,
https://doi.org/10.1007/978-1-0716-1229-3_13,

[4, 5]. Indeed, 5fC alters the structure of DNA double helix [6], increases DNA flexibility [7], recruits specific proteins including transcriptional regulators, DNA-repair factors, and chromatin regulators [8], yields reversible-covalent Schiff base linkage with histone residues [9–12], shows distinct genomic profile from that of 5mC, 5hmC or 5caC at single-base resolution [13–15], and accumulates at functionally important genomic regions, thus highlighting its regulatory role in chromatin remodeling, gene expression, cellular differentiation, and embryonic development.

5fU, the modified thymidine counterpart of 5fC, is a well-known genotoxic lesion in most cells and tissues [16]. Its abundance is comparable to that of 5fC, nearly 1–10 moieties per 10^6 nucleobases [17–22]. Exposure of thymidine to UV light, ionizing radiation, reactive oxygen species (ROS), Fenton reagents or particular enzymes raises 5fU level [23]. 5fU and 5fC have little difference in structure and both contain a typical 5-formyl group, but the latter one is less reactive due to the presence of the intramolecular H-bond between the exocyclic amino group $NH_2(4)$ and the carbonyl oxygen at C5 [24]. Therefore, it has always been a daunting task that how to distinguish 5fC from 5fU and how to develop facile, robust, and reliable detection methods for 5fC in DNA, so as to gain a deeper insight of epigenetic functions and relevant underlying mechanism of 5fC.

Owing to inherent sensitivity and selectivity, LC-MS^n has been widely used in the analysis of rare nucleosides. However, quantifications of 5fC is challenging due to its extremely low abundance in vivo (for example, ~0.002% of all cytosine residues in mouse ESCs) [25, 26], low ionization efficiency as well as the multiple interferences from the sample matrix. Additional chemical derivatization with easily ionized tertiary or quaternary ammonia has been proved to be a promising strategy to improve the detection performance of 5fC in MS analysis [27, 28]. Comparison with the time-consuming, expensive and isotope-labeled internal standard needed LC-MS^n, fluorescence-based detection methods are more economical, faster, and easier to operate. 5fC can be fluorescently switch on by amine [29], hydroxylamine [30–32], and hydrazine [33, 34] derivatives. Nevertheless, such Schiff base type reactions have their inherent drawbacks, for example, imine formation (especially oxime linkage) at neutral pH suffer from poor reactivity and sometimes need catalysts. Besides, the formed C=N is susceptible to hydrolysis over time. Moreover, as mentioned above, 5fU is more reactive than 5fC, so these amine derivatives preferentially reacting with 5fU, thereby seriously disturbing the determination of 5fC [35]. In 2016, Höbartner and co-workers first explored an aldol-type condensation toward 5fC/5fU with 2,3,3-trimethylindole derivatives [36], in which a breakthrough C=C was constructed. More importantly, based on the inherently distinct electronic properties of 5fC and 5fU, the corresponding hemicyanine (Hcy)-like fluorescent nucleosides have distinct excitation/emission maxima, allowing the two to be distinguished. Similar to Friedländer synthesis to form quinoline derivatives, Yi [37, 38] and Zhou [39, 40] et al. reported a class of -CH_2- reagents that can highly selectively condense with 5fC bearing a characteristic 2-aminobenzaldehyde structure to yield fluorescent intramolecular-cyclized nucleobases, whereas 5fU and abasic sites (AP) without this structure do not interfere. They achieved 5fC-specific recognition through a "dual sites" targeting strategy, however, these -CH_2- reagents bearing two electron-withdrawing groups are not reactive enough as they take a long time (10–24 h) to completely label 5fC. In this chapter, we provide a detailed description of the photo-assisted triple domino reaction strategy for rapid qualitative and quantitative analysis of 5fC with excellent sensitivity and selectivity that we recently developed [41].

We screened a lot of phosphorus ylides (P-ylides) and found the commercially available (Triphenylphosphoranylidene)acetonitrile, termed **YC-CN**, is promising. Of course, both 5fC and 5fU can form a stable C=C with the highly reactive **YC-CN** via a Wittig olefination pathway. Due to the *E*-preference of **YC-CN**, the major products are *trans*-isomers, but they are not fluorescent. Application of UV irradiation will effectively convert *trans*-**5fC-CN** to *cis*-**5fC-CN**, in which -CN is closer to exocyclic 4-NH_2 and can further undergo intramolecular cycloaddition to afford a high fluorescent **5fC-CN-Close**. The nucleoside-derived fluorophore possesses an extremely high quantum yield ($\Phi = 0.87$), excellent photostability, and good water solubility. However, **5fU-CN** cannot undergo such conversion due to the lack of vital 4-NH_2 element, which making it feasible for us to distinguish 5fC from 5fU and accurately quantify 5fC with high selectivity (Fig. 1). It is worth noting that the yield of Wittig reaction and photocyclization are almost 100%. Furthermore, to the best of our knowledge, the time-consuming in our fluorescent labeling strategy is the shortest.

This chapter aims at providing guidance on how to optimize the derivatization conditions of 5fC (including reaction temperature, reaction time, amount of **YC-CN** and irradiation time), and how to analyze 5fC content in deoxynucleoside mixtures or enzyme-digested DNA extracted from cells, blood, tissues, etc.

2 Materials

2.1 Reagents and Solutions

- 5fC: **dissolve in DMSO at 0.5 mg/100 μL or 10 mM** to prepare the stock solution.
- 2′-deoxynucleoside stock solution (A, G, C, T, U, and 5mC are purchased from Aladdin; 5hmC, 5hmU, and 5fU are self-synthesized): **dissolve in DMSO at 10 mM, respectively.**

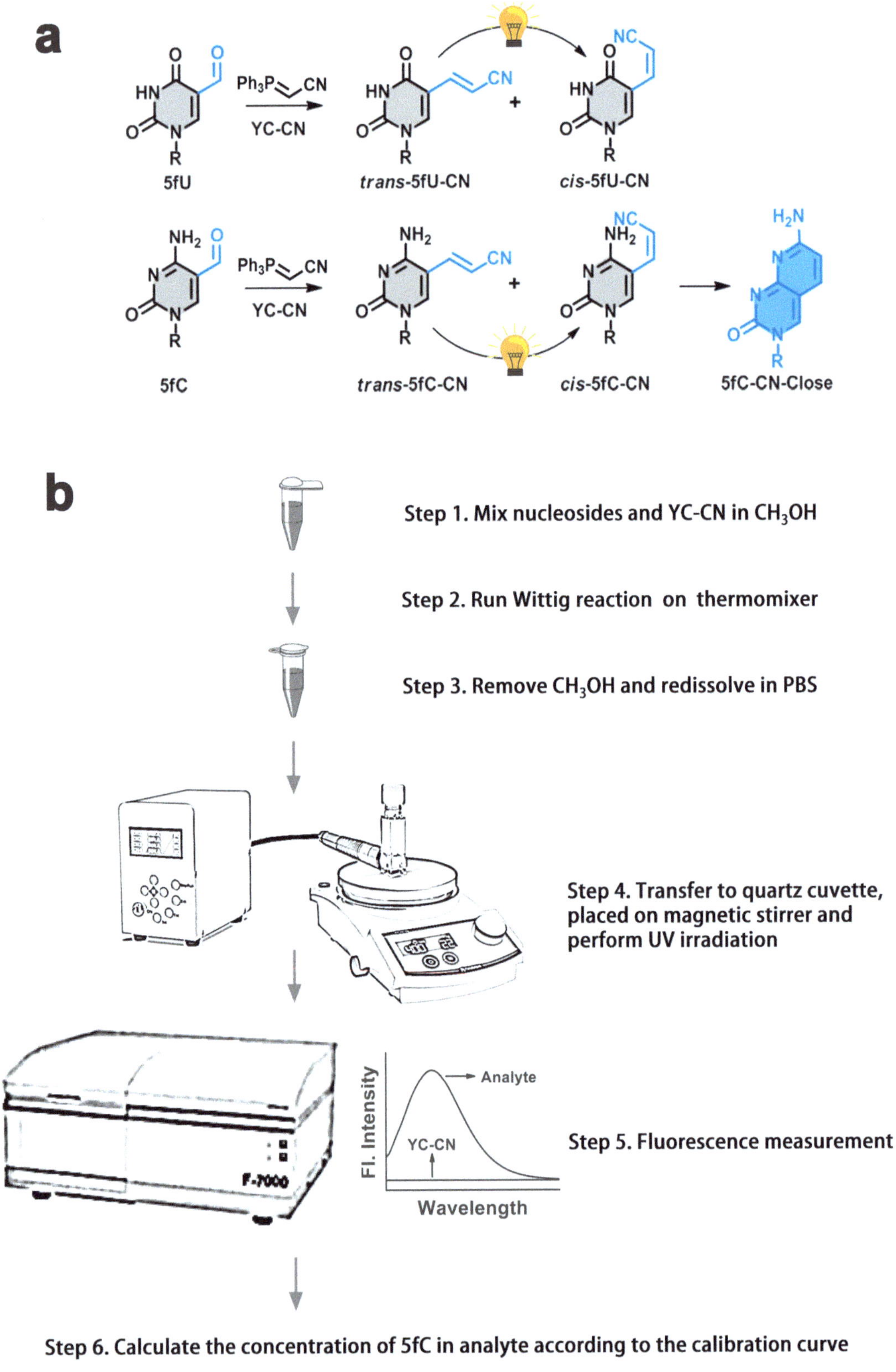

Fig. 1 Application of photo-assisted domino reaction for 5fC analysis. (**a**) Selectively fluorescence labeling of 5fC with Wittig reagent (**YC-CN**). Reproduced from ref. [41] with permission from American Chemical

- **YC-CN** (Aladdin, cat. no. T162823): **dissolve in water at 0.1 mM to prepare the** stock solution.
- **Phosphate-buffered saline (PBS): 8.00 g NaCl (0.137 M), 0.20 g KCl (2.7 mM), 0.24 g KH_2PO_4 (1.4 mM), 1.44 g Na_2HPO_4 (0.01 M) are dissolved in 1 L H_2O.** The pH of the solution was adjusted to 7.4 with HCl.
- ***Trans*-5fC-CN (self-synthesized): dissolve in DMSO at 100 mM to prepare the stock solution.**
- **5fC-CN-Close (self-synthesized): dissolve in PBS at 0–10 μM to prepare the standard solution for calibration curve.**

2.2 Equipment

- 1.5-mL microcentrifuge tubes.
- Vortex Mixer.
- Eppendorf ThermoMixer.
- Microcentrifuge (ThermoFisher Scientific, cat. no. 75002430).
- Syringe filters (0.45 μm).
- High-performance liquid chromatography (HPLC): Waters Alliance HPLC system equipped with a Waters e2695 Separations Module and a 2998 Photodiode Array (PDA) Detector (Waters, Milford, MA, USA); an Ultimate® XB-C18 column (5 μm, 300 Å, 4.6 × 250 mm, Welch, China) was used for separation.
- 280 nm UV LED spot curing system (light power 0.4 mW, spot size ~0.785 cm^2).
- UV-visible Spectrophotometer (TU-1901, Persee, China).
- Fluorescence Spectrophotometer (F-7000, Hitachi, Japan).
- Quartz cuvette for photocyclization (standard rectangular, H × W × D = 48 mm × 12.5 mm × 12.5 mm, pathlength 10 × 10 mm, chamber volume 3.5 mL, with PTFE stopper).
- Magnetic Stirrer (Heodolph, Hei-Tec).
- Pipets and pipet tips (we recommend pipet tips with aerosol barriers for preventing cross-contamination).
- Nitrogen Dryer.

Fig. 1 (continued) Society. (**b**) Overview of the 6 steps workflow to qualitative and quantitative detection of 5fC in nucleoside mixtures

3 Methods

3.1 Optimization of Derivatization Conditions for 5fC by YC-CN (see Note 1)

3.1.1 Optimization of Wittig Olefination Conditions

We performed the condition optimization experiments of Wittig olefination between 5fC and **YC-CN** using 0.5 mg 5fC (~2.612 mM) as substrate, which is sufficient to analyze the 5fC levels in almost all sample DNA as they are believed to exist in low abundance. However, it is also acceptable to appropriately reduce the substrate concentration during this optimization process to obtain a milder reaction condition, such as lower reaction temperature, shorter reaction time, less probe amount, etc.

All HPLC analysis were performed on a Waters Alliance HPLC system equipped with a Waters e2695 Separations Module and a 2998 Photodiode Array (PDA) Detector. The Ultimate® XB-C18 column (5 μm, 300 Å, 4.6 × 250 mm, Welch, China), maintained at 35 °C, was used for separation. The injection volume was 10 μL and the detector wavelength was fixed at 260 nm. Water (solvent A) and CH_3OH (solvent B) were employed as the mobile phase at a flow rate of 1 mL/min. Gradient elution program: 10-10-100-100%B /0-15-20-25 min.

Reaction temperature

1. Add 100 μL 5fC stock solution (**0.5 mg/100 μL in DMSO**), 5.9 mg **YC-CN** (10 eq.), and 650 μL CH_3OH into 1.5 mL microcentrifuge tubes. Mix thoroughly by vortexing.
2. Set up the Wittig reactions in a thermomixer (Eppendorf, 1400 rpm) at different temperatures ranging from room temperature to 80 °C (for example, 30 °C, 40 °C, 50 °C, 60 °C, 70 °C, and 80 °C) for 0.5 h.
3. Add an equal volume of water and vortex for 30 s to quench the reactions. Briefly centrifuge to remove the droplets from inside the lid.
4. Filter through a 0.45 μm microfiltration membrane and then subject to HPLC analysis.
5. Select the optimal reaction temperature for subsequent optimization. (Our results demonstrated that the maximum peak area of **5fC-CN** can be achieved at 60 °C.)

Reaction time

1. Add 100 μL 5fC stock solution (**0.5 mg/100 μL in DMSO**), 5.9 mg **YC-CN** (10 eq.), and 650 μL CH_3OH into 1.5 mL microcentrifuge tubes. Mix thoroughly by vortexing.
2. Set up the Wittig reactions in a thermomixer (Eppendorf, 1400 rpm) at 60 °C for different times ranging from 0.5 to 3 h (for example, 0.5, 1, 1.5, 2, 2.5, and 3 h).

3. Add an equal volume of water and vortex for 30 s to quench the reactions. Briefly centrifuge to remove the droplets from inside the lid.
4. Filter through a 0.45 μm microfiltration membrane and then subject to HPLC analysis.
5. Select the optimal reaction time for subsequent optimization. (Our results show that 0.5 h, or at most 1 h, is sufficient.)

Amounts of **YC-CN**

1. Add 100 μL 5fC stock solution (**0.5 mg/100 μL in DMSO**), different amounts of **YC-CN** ranging from 2 to 150 eq. (for example, 1.2 mg (2 eq.), 3.0 mg (5 eq.), 5.9 mg (10 eq.), 11.8 mg (20 eq.), 17.7 mg (30 eq.), 23.6 mg (40 eq.), 29.5 mg (50 eq.), 44.3 mg (75 eq.), 59.0 mg (100 eq.), 88.5 mg (150 eq.)) and 650 μL CH_3OH into 1.5 mL microcentrifuge tubes. Mix thoroughly by vortexing.
2. Set up the Wittig reactions in a thermomixer (Eppendorf, 1400 rpm) at 60 °C for 1 h.
3. Add an equal volume of water and vortex for 30 s to quench the reactions. Briefly centrifuge to remove the droplets from inside the lid.
4. Filter through a 0.45 μm microfiltration membrane and then subject to HPLC analysis.
5. Choose the optimal amounts of **YC-CN** for subsequent optimization. (Our results demonstrated that 20–100 eq. **YC-CN** can achieve ~100% conversion of 5fC.)

3.1.2 Optimization of Photocatalytic Reaction Time

When optimizing the irradiation time, we performed a photocatalytic reaction of 0.1 mM *trans*-**5fC-CN** in PBS using a 280 nm UV LED spot curing system at room temperature (*see* **Note 2**). The synthetic route of *trans*-**5fC-CN** and the optimization steps of photocatalytic reaction time are described below (Fig. 2).

5fC (66 mg, 0.2587 mmol) was added to a solution of **YC-CN** (156 mg, 0.5175 mmol) in 5 mL methanol. The reaction mixture was kept stirring at 37 °C overnight. Then, the solvent was evaporated in vacuo and the crude product was purified by silica gel column chromatography (CH_2Cl_2: Methanol = 10:1, v/v) to afford *trans*-**5fC-CN** and a little **5fC-CN-Close**.

trans-**5fC-CN**: ^{1}H NMR (400 MHz, DMSO-d_6) δ 8.57 (s, 1H), 7.52 (d, J = 16.3 Hz, 1H), 6.04 (t, J = 5.9 Hz, 1H), 5.82 (d, J = 16.3 Hz, 1H), 5.25 (t, J = 4.6 Hz, 1H), 5.22 (d, J = 4.4 Hz, 1H), 4.19 (dt, J = 9.5, 4.6 Hz, 1H), 3.78 (dd, J = 7.0, 3.2 Hz, 1H), 3.72–3.50 (m, 2H), 2.24–1.96 (m, 2H); ^{13}C NMR (101 MHz, DMSO-*d6*) δ 163.0, 154.1, 143.1, 141.8, 119.5, 101.3, 92.6, 87.8, 86.1, 69.5, 60.8, 41.4; HRMS (ESI) *m/z* calcd for $[M + H]^+$: 301.0913, found: 301.0769.

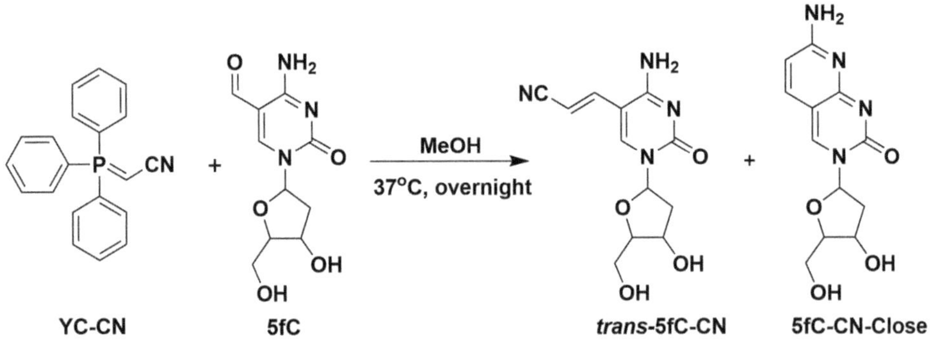

Fig. 2 The synthetic route of *trans*-**5fC-CN** and **5fC-CN-Close**

5fC-CN-Close: ^{1}H NMR (400 MHz, DMSO-d_6) δ 7.31 (d, J = 8.4 Hz, 1H), 6.41 (s, 2H), 6.09 (d, J = 8.3 Hz, 1H), 5.88 (s, 1H), 5.86 (s, 1H), 5.12 (d, J = 4.3 Hz, 1H), 4.26 (s, 1H), 4.22–4.14 (m, 1H), 3.98–3.75 (m, 2H), 2.44–2.07 (m, 2H); ^{13}C NMR (101 MHz, DMSO-*d6*) δ 160.1, 150.7, 146.8, 139.0, 102.6, 99.4, 90.4, 88.1, 84.5, 72.0, 71.4, 46.4; HRMS (ESI) *m/z* calcd for $[M + Na]^+$: 279.1093, found: 279.0927.

1. Dissolve *trans*-**5fC-CN** in water to prepare a 0.1 mM solution. Transfer 750 μL of the solution into a standard 3.5 mL quartz cuvette with PTFE stopper.
2. Place the reaction on a Magnetic Stirrer. Start stirring and UV irradiation (Fig. 1b).
3. Aliquots (50 μL) are removed from the reaction mixture at different time points (5 min, 10 min, 20 min, 30 min, 45 min, 1 h, 1.5 h, 2 h, 2.5 h, 3 h, 3.5 h, 4 h, and 4.5 h) and diluted to 3 mL PBS to record UV–vis absorption and fluorescence spectra.
4. Plot of fluorescence intensity at 410 nm versus time will help to find the optimal irradiation time. Meanwhile, UV−vis spectra are also good reference, in which the characteristic absorption band centered at 273 nm belonged to *trans*-**5fC−CN** dropped gradually and a weak shoulder peak at 350 nm appeared as the irradiation time prolonged.

Taken together, the optimized derivatization conditions for 5fC by **YC-CN** are at 60 °C for 1 h with 20–100 eq. **YC-CN**, followed by 1.5 h irradiation of 280 nm UV. Under the optimized conditions, 5fC was completely converted to **5fC−CN-Close**. However, other derivatization conditions are also applicable, as long as the yield of **5fC−CN-Close** is sufficiently high.

3.2 Verify the Specificity of YC−CN for 5fC

Before quantifying 5fC in actual samples, it is necessary to verify the specificity of this method, which may be influenced by the varied environments, reagents, equipment, etc.

1. Add 1 μL 2′-deoxynucleoside stock solution (10 mM in DMSO, including A, G, C, T, U, 5mC, 5hmC, 5fC, 5hmU, 5fU), 2μL **YC-CN** (100 mM in DMSO, freshly prepared), and 100 μL CH_3OH into 1.5 mL microcentrifuge tubes, respectively. Carefully pipet up and down several times to mix the reaction mixtures thoroughly.
2. Set up the Wittig reactions in a thermomixer (Eppendorf, 1400 rpm) at 60 °C for 1 h.
3. Open lids of each microcentrifuge tubes and the solvent CH_3OH was then volatilized with nitrogen gas at 37 °C. Redissolve in 1 mL PBS and vortex for 1 min to ensure thoroughly mixing. Briefly centrifuge to remove the droplets from inside the lid.
4. Transfer 750 μL of solution from **step 3** into the corresponding 3.5 mL quartz cuvettes placed on magnetic stirrers. Start stirring and UV irradiation for 1.5 h.
5. Scan fluorescence spectra upon excitation at 345 nm in turn and record the fluorescence intensity at 410 nm.

Normally, only 5fC triggers a remarkable fluorescent enhancement at 410 nm, whereas other potential interferences, even 5fU, induce no obvious spectral changes of **YC-CN**. Make sure there is negligible interference before sample analysis.

3.3 Analyze 5fC Content in DNA Samples

We carry out 5fC quantification using dried deoxynucleosides as starting materials. If not, necessary pretreatment steps, such as DNA extraction, enzymatic digestion, purification, and drying are required. Extracting of high-quality genome DNA from cells, blood, and tissues can be easily achieved using commercially available DNA extraction kits. The detailed methods for DNA digestion have been described in a number of previous publications [42].

1. The dried deoxynucleoside mixture in 1.5 mL microcentrifuge tubes were added 2 μL **YC-CN** (100 mM in DMSO, freshly prepared) and 100 μL CH_3OH (*see* **Note 3**). Mix thoroughly by vortexing and make sure the deoxynucleosides are completely dissolved. Briefly centrifuge to remove the droplets from inside the lid.
2. Set up the Wittig reactions in a thermomixer (Eppendorf, 1400 rpm) at 60 °C for 1 h.
3. Open lids of each microcentrifuge tubes and the solvent CH_3OH was then volatilized with nitrogen gas at 37 °C (*see* **Note 4**). Redissolve in 1 mL PBS and vortex for 1 min to ensure thoroughly mixing (*see* **Note 5**). Briefly centrifuge to remove the droplets from inside the lid.

4. Transfer 750 μL of solution from **step 3** into the corresponding 3.5 mL quartz cuvettes placed on magnetic stirrers (*see* **Note 6**). Start stirring and UV irradiation for 1.5 h.
5. Scan fluorescence spectra upon excitation at 345 nm in turn and record the fluorescence intensity at 410 nm.
6. Calculate the concentration of 5fC according to the calibration curve and the average fluorescence intensity of the sample measured in triplicate (*see* **Note 7**).

4 Notes

1. Optimization steps of derivatization conditions for 5fC by **YC-CN** are optional. For example, if a different UV instrument was utilized for photocatalytic reaction, the irradiation time should be carefully reinvestigated. We have provided an optimized condition for 5fC conversion with almost 100% yield, which can be used directly for sample analysis if the equipment and reagent sources are similar to those described in this protocol.
2. We have screened the light source for photocyclization of *trans*-**5fC-CN**, including 280 nm LED, 310 nm LED, 365 nm LED, and high-pressure mercury lamp (the strongest emission peak is 365 nm). Our results demonstrated that *trans*-**5fC-CN** was almost completely deteriorated after irradiation by the latter two, while the former two can effectively realize *trans*-**5fC-CN** to **5fC-CN-Close** conversion and 280 nm LED is better than 310 nm LED. The best performance of the 280 nm LED can also be explained by the UV-vis absorption spectrum of *trans*-**5fC-CN** and **5fC-CN-Close**. The absorption peaks of *trans*-**5fC-CN** are 272 nm, followed by 306 nm, and extremely weak at 365 nm [41]. Of course, the *trans*-to-*cis* isomerization capabilities of the above three light sources with different wavelengths also follow this sequence, which is consistent with our experimental results. Moreover, the cyclization product, **5fC-CN-Close**, has a distinct absorption peak at nearly 310 nm; therefore, continuous 310 nm UV irradiation may accelerate its decomposition. We finally select a 280 nm UV LED spot curing system for photocyclization. However, other instruments emitting ~280 nm UV light are also applicable, and the stronger the light, the shorter the radiation time.
3. The solvent of Wittig reaction between 5fC (or 5fC-containing samples) and **YC-CN** are not limited to CH_3OH, but they should satisfy the following characteristics: (1) good stability; (2) good solubility of **YC-CN**; (3) good solubility of 5fC and

5fC-containing deoxynucleoside mixtures; (4) low boiling point, easy to volatilize after Wittig olefination. The solvent volume of Wittig reaction also is not limited to "100 μL," but a larger volume may lead to longer reaction time and longer drying time. Notably, water cannot be introduced during Wittig reaction because **YC-CN** is sensitive to water and will completely decompose in the presence of water, resulting in chemical labeling to failure.

4. After Wittig reaction, the solvent volatilization step cannot be omitted. Because **5fC-CN-Close** exhibits significant decrease in fluorescence intensity and a blue shift of ~30 nm in the maximum emission wavelength in most organic solvents, such as methanol (CH_3OH), ethanol (EtOH), N, N-dimethylformamide (DMF), N,N-dimethylacetamide, dimethyl sulfoxide (DMSO), etc. Large amounts of residual solvent will make sample analysis unreliable.
5. In **step 3** of Subheading 3.3, the volume of PBS is not limited to "1 mL." When analyzing DNA samples with low abundance of 5fC, it is recommended to use a small volume of PBS for redissolution to obtain concentrated analytes, which is an alternative strategy to harvest enhanced fluorescence.
6. In **step 4** of Subheading 3.3, the volume transferred to quartz cuvette is not limited, but make sure the reaction mixture is completely exposed to UV irradiation.
7. Calibration curve for 5fC analysis was constructed before each analysis using a standard solution of **5fC-CN-Close**. Good linearity within the concentration range of 0–10 μM was obtained ($R^2 > 0.99$). The synthesis of **5fC-CN-Close** was described in Subheading 3.1.2 with limited yield, alternatively, UV irradiation of *trans*-**5fC-CN** will afford **5fC-CN-Close** more effectively.

Acknowledgements

This work was financially supported by National Natural Science Foundation of China (No.: 21572147 and 21472131). We also thank the Comprehensive Training Platform of Specialized Laboratory, College of Chemistry, Sichuan University for sample analysis.

References

1. Greenberg MVC, Bourc'his D (2019) The diverse roles of DNA methylation in mammalian development and disease. Nat Rev Mol Cell Biol 20:590–607
2. Deckard CE 3rd, Banerjee DR, Sczepanski JT (2019) Chromatin structure and the pioneering transcription factor FOXA1 regulate TDG-mediated removal of 5-formylcytosine

from DNA. J Am Chem Soc 141:14110–14114
3. Carell T, Kurz MQ, Muller M, Rossa M, Spada F (2018) Non-canonical bases in the genome: the regulatory information layer in DNA. Angew Chem Int Ed 57:4296–4312
4. Su M, Kirchner A, Stazzoni S, Mueller M, Wagner M, Schroeder A, Carell T (2016) 5-formylcytosine could be a semipermanent base in specific genome sites. Angew Chem Int Ed 55:11797–11800
5. Bachman M, Uribe-Lewis S, Yang X, Burgess HE, Iurlaro M, Reik W, Murrell A, Balasubramanian S (2015) 5-formylcytosine can be a stable DNA modification in mammals. Nat Chem Biol 11:555–557
6. Raiber EA, Murat P, Chirgadze DY, Beraldi D, Luisi BF, Balasubramanian S (2015) 5-formylcytosine alters the structure of the DNA double helix. Nat Struct Mol Biol 22:44–49
7. Ngo TT, Yoo J, Dai Q, Zhang Q, He C, Aksimentiev A, Ha T (2016) Effects of cytosine modifications on DNA flexibility and nucleosome mechanical stability. Nat Commun 7:10813
8. Iurlaro M, Ficz G, Oxley D, Raiber EA, Bachman M, Booth MJ, Andrews S, Balasubramanian S, Reik W (2013) A screen for hydroxymethylcytosine and formylcytosine binding proteins suggests functions in transcription and chromatin regulation. Genome Biol 14:R119
9. Raiber E-A, Portella G, Cuesta SM, Hardisty R, Murat P, Li Z, Iurlaro M, Dean W, Spindel J, Beraldi D, Liu Z, Dawson MA, Reik W, Balasubramanian S (2018) 5-formylcytosine organizes nucleosomes and forms Schiff base interactions with histones in mouse embryonic stem cells. Nat Chem 10:1258–1266
10. Ji S, Fu I, Naldiga S, Shao H, Basu AK, Broyde S, Tretyakova NY (2018) 5-formylcytosine mediated DNA-protein cross-links block DNA replication and induce mutations in human cells. Nucleic Acids Res 46:6455–6469
11. Li F, Zhang Y, Bai J, Greenberg MM, Xi Z, Zhou C (2017) 5-formylcytosine yields DNA-protein cross-links in nucleosome core particles. J Am Chem Soc 139:10617–10620
12. Ji S, Shao H, Han Q, Seiler CL, Tretyakova NY (2017) Reversible DNA-protein cross-linking at epigenetic DNA Marks. Angew Chem Int Ed 56:14130–14134
13. Song C-X, Szulwach KE, Dai Q, Fu Y, Mao S-Q, Lin L, Street C, Li Y, Poidevin M, Wu H, Gao J, Liu P, Li L, Xu G-L, Jin P, He C (2013) Genome-wide profiling of 5-formylcytosine reveals its roles in epigenetic priming. Cell 153:678–691
14. Raiber E-A, Beraldi D, Ficz G, Burgess HE, Branco MR, Murat P, Oxley D, Booth MJ, Reik W, Balasubramanian S (2012) Genome-wide distribution of 5-formylcytosine in embryonic stem cells is associated with transcription and depends on thymine DNA glycosylase. Genome Biol 13:R69
15. Iurlaro M, McInroy GR, Burgess HE, Dean W, Raiber E-A, Bachman M, Beraldi D, Balasubramanian S, Reik W (2016) In vivo genome-wide profiling reveals a tissue-specific role for 5-formylcytosine. Genome Biol 17:141
16. Bjelland S, Anensen H, Knevelsrud I, Seeberg E (2001) Cellular effects of 5-formyluracil in DNA. Mutat Res 486:147–154
17. Jiang H-P, Liu T, Guo N, Yu L, Yuan B-F, Feng Y-Q (2017) Determination of formylated DNA and RNA by chemical labeling combined with mass spectrometry analysis. Anal Chim Acta 981:1–10
18. Zhang H-Y, Xiong J, Qi B-L, Feng Y-Q, Yuan B-F (2016) The existence of 5-hydroxymethylcytosine and 5-formylcytosine in both DNA and RNA in mammals. Chem Commun 52:737–740
19. Huang W, Lan M-D, Qi C-B, Zheng S-J, Wei S-Z, Yuan B-F, Feng Y-Q (2016) Formation and determination of the oxidation products of 5-methylcytosine in RNA. Chem Sci 7:5495–5502
20. Gackowski D, Starczak M, Zarakowska E, Modrzejewska M, Szpila A, Banaszkiewicz Z, Olinski R (2016) Accurate, direct, and high-throughput analyses of a broad spectrum of endogenously generated DNA base modifications with isotope-dilution two-dimensional ultraperformance liquid chromatography with tandem mass spectrometry: possible clinical implication. Anal Chem 88:12128–12136
21. Bauer NC, Corbett AH, Doetsch PW (2015) The current state of eukaryotic DNA base damage and repair. Nucleic Acids Res 43:10083–10101
22. Pfaffeneder T, Spada F, Wagner M, Brandmayr C, Laube SK, Eisen D, Truss M, Steinbacher J, Hackner B, Kotljarova O, Schuermann D, Michalakis S, Kosmatchev O, Schiesser S, Steigenberger B, Raddaoui N, Kashiwazaki G, Mueller U, Spruijt CG, Vermeulen M, Leonhardt H, Schaer P, Mueller M, Carell T (2014) Tet oxidizes thymine to 5-hydroxymethyluracil in mouse

embryonic stem cell DNA. Nat Chem Biol 10:574–581

23. Liu C, Zou G, Peng S, Wang Y, Yang W, Wu F, Jiang Z, Zhang X, Zhou X (2018) 5-formyluracil as a multifunctional building block in biosensor designs. Angew Chem Int Ed 57:9689–9693
24. Munzel M, Lischke U, Stathis D, Pfaffeneder T, Gnerlich FA, Deiml CA, Koch SC, Karaghiosoff K, Carell T (2011) Improved synthesis and mutagenicity of oligonucleotides containing 5-hydroxymethylcytosine, 5-formylcytosine and 5-carboxylcytosine. Chem Eur J 17:13782–13788
25. Wagner M, Steinbacher J, Kraus TFJ, Michalakis S, Hackner B, Pfaffeneder T, Perera A, Mueller M, Giese A, Kretzschmar HA, Carell T (2015) Age-dependent levels of 5-methyl-, 5-hydroxymethyl-, and 5-formylcytosine in human and mouse brain tissues. Angew Chem Int Ed 54:12511–12514
26. Pfaffeneder T, Hackner B, Truss M, Muenzel M, Mueller M, Deiml CA, Hagemeier C, Carell T (2011) The discovery of 5-formylcytosine in embryonic stem cell DNA. Angew Chem Int Ed 50:7008–7012
27. Zhang YF, Qi CB, Yuan BF, Feng YQ (2019) Determination of cytidine modifications in human urine by liquid chromatography-mass spectrometry analysis. Anal Chim Acta 1081:103–111
28. Qi C, Jiang H, Xiong J, Yuan B, Feng Y (2019) On-line trapping/capillary hydrophilic-interaction liquid chromatography/mass spectrometry for sensitive determination of RNA modifications from human blood. Chin Chem Lett 30:553–557
29. Hu J, Xing X, Xu X, Wu F, Guo P, Yan S, Xu Z, Xu J, Weng X, Zhou X (2013) Selective chemical labelling of 5-formylcytosine in DNA by fluorescent dyes. Chem Eur J 19:5836–5840
30. Wang Y, Liu C, Yang W, Zou G, Zhang X, Wu F, Yu S, Luo X, Zhou X (2018) Naphthalimide derivatives as multifunctional molecules for detecting 5-formylpyrimidine by both PAGE analysis and dot-blot assays. Chem Commun 54:1497–1500
31. Hong T, Wang T, Guo P, Xing X, Ding F, Chen Y, Wu J, Ma J, Wu F, Zhou X (2013) Fluorescent strategy based on cationic conjugated polymer fluorescence resonance energy transfer for the quantification of 5-(hydroxymethyl)cytosine in genomic DNA. Anal Chem 85:10797–10802
32. Guo P, Yan S, Hu J, Xing X, Wang C, Xu X, Qiu X, Ma W, Lu C, Weng X, Zhou X (2013) Selective detection of 5-formyl-2′-deoxycytidine in DNA using a fluorogenic hydroxylamine reagent. Org Lett 15:3266–3269
33. Xu L, Chen Y-C, Nakajima S, Chong J, Wang L, Lan L, Zhang C, Wang D (2014) A chemical probe targets DNA 5-formylcytosine sites and inhibits TDG excision, polymerases bypass, and gene expression. Chem Sci 5:567–574
34. Xu L, Chen Y-C, Chong J, Fin A, McCoy LS, Xu J, Zhang C, Wang D (2014) Pyrene-based quantitative detection of the 5-formylcytosine loci symmetry in the CpG duplex content during TET-dependent demethylation. Angew Chem Int Ed 53:11223–11227
35. Liu C, Luo X, Chen Y, Wu F, Yang W, Wang Y, Zhang X, Zou G, Zhou X (2018) Selective labeling aldehydes in DNA. Anal Chem 90:14616–14621. https://doi.org/10.1021/acs.analchem.8b04822
36. Samanta B, Seikowski J, Hoebartner C (2016) Fluorogenic labeling of 5-formylpyrimidine nucleotides in DNA and RNA. Angew Chem Int Ed 55:1912–1916
37. Zhu C, Gao Y, Guo H, Xia B, Song J, Wu X, Zeng H, Kee K, Tang F, Yi C (2017) Single-cell 5-formylcytosine landscapes of mammalian early embryos and ESCs at single-base resolution. Cell Stem Cell 20:720–731
38. Wu H, Wu X, Zhang Y (2016) Base-resolution profiling of active DNA demethylation using MAB-seq and caMAB-seq. Nat Protoc 11:1081–1100
39. Liu C, Wang Y, Yang W, Wu F, Zeng W, Chen Z, Huang J, Zou G, Zhang X, Wang S, Weng X, Wu Z, Zhou Y, Zhou X (2017) Fluorogenic labeling and single-base resolution analysis of 5-formylcytosine in DNA. Chem Sci 8:7443–7447
40. Wang Y, Liu C, Zhang X, Yang W, Wu F, Zou G, Weng X, Zhou X (2018) Gene specific-loci quantitative and single-base resolution analysis of 5-formylcytosine by compound-mediated polymerase chain reaction. Chem Sci 9:3723–3728
41. Zhou Q, Li K, Li LL, Yu KK, Zhang H, Shi L, Chen H, Yu XQ (2019) Combining Wittig olefination with photoassisted Domino reaction to distinguish 5-formylcytosine from 5-formyluracil. Anal Chem 91:9366–9370
42. Tang Y, Xiong J, Jiang HP, Zheng SJ, Feng YQ, Yuan BF (2014) Determination of oxidation products of 5-methylcytosine in plants by chemical derivatization coupled with liquid chromatography/tandem mass spectrometry analysis. Anal Chem 86:7764–7772

Chapter 14

Detection of 5-Formyluracil and 5-Formylcytosine in DNA by Fluorescence Labeling

Chaoxing Liu

Abstract

5-Formyluracil (5fU) and 5-formylcytosine (5fC) which are widely present in human genomic DNAs play significant roles in epigenetic functions and have attracted widespread attention in many related fields. Therefore, creating highly effective, selective, and easy-operating detection methods for these important natural existing DNA modifications is important not only to understand the fundamentals of physiological regulation, but also serve as the basis for the next generation of therapeutics used to improve human's health. Within last decades, various methods have been developed to qualitatively and quantitatively detect these modifications. We describe in detail the protocols of fluorescence labeling methods for detection of 5fU and 5fC in DNA. The highly selective fluorescence "switch-on" specificity towards 5fU or 5fC separately enabled a high signal-to-noise ratio in qualitatively and quantitatively detecting 5fU or 5fC and it is not affected by the presence of other DNA modifications which also bear formyl groups. These protocols offer solutions to problems related to fast, convenient, cost-efficient, and easy-operating detection of 5fU or 5fC in complex samples.

Keywords 5-Formyluracil, 5-Formylcytosine, Fluorescence labeling, DNA modifications, Fluorescent probes

1 Introduction

DNA formyl-modified structures such as 5-formyluracil (5fU), 5-formylcytosine (5fC), and abasic sites (AP) are widely existed in human genomic DNAs (Fig. 1) [1–3]. 5fU can be generated from the oxidation via gamma-ray damage, ultraviolet radiation, reactive oxygen attack, thymidine-7 hydroxylase oxidation, and Fenton reagent treatment of thymine or via AID/APOBEC enzymes by replacing 4′-site amino group from 5fC (Fig. 2) [4–7]. 5fU is not only considered a genetic damage biomarker with strong genotoxicity but also serves as a potential cancer biomarker for thyroid cancer [8, 9]. 5fC can also be obtained from 5-methylcytosine by Tet (ten–eleven translocation) active oxidization or ROS passive oxidization (Fig. 2) [10, 11]. 5fC plays a significant part in

Bi-Feng Yuan (ed.), *DNA Modification Detection Methods*, Springer Protocols Handbooks,
https://doi.org/10.1007/978-1-0716-1229-3_14,

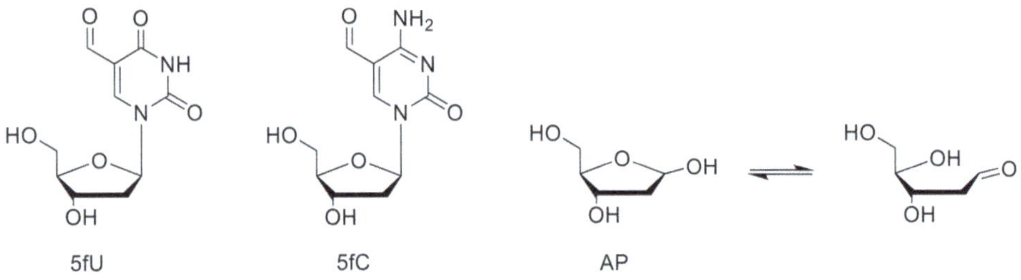

Fig. 1 Structures of DNA formyl-modified nucleobase. 5-formyluracil (5fU), 5-formylcytosine (5fC), and abasic sites (AP)

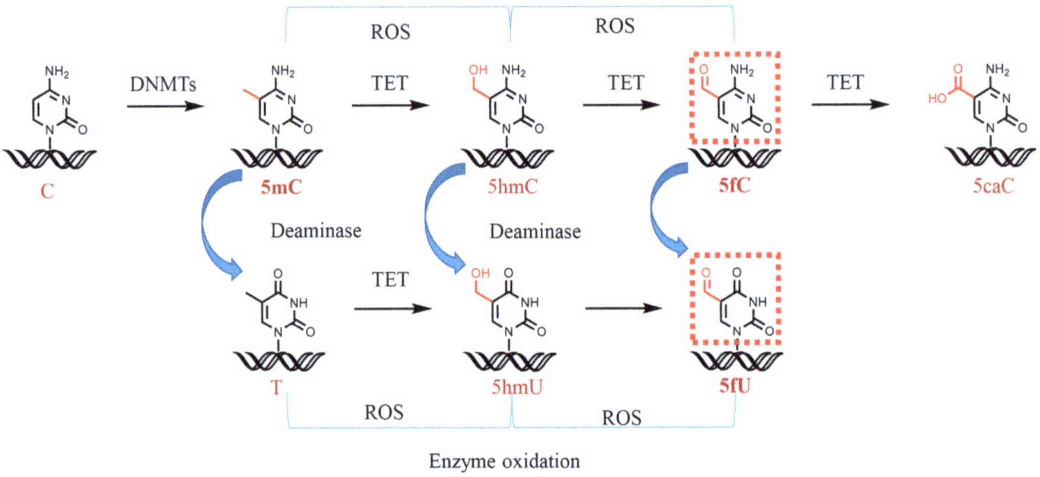

Fig. 2 The correlation of 5fU and 5fC with canonical nucleobases

epigenetic functions due to its stability in various genomes and associated with alterations in DNA structures, gene regulation, cell differentiation, and diseases [12–14].

Research on 5fU and 5fC has paved the way for completely understanding of genetic and epigenetic regulation and advance the next generation of therapeutics based on DNA epigenetics [15–20]. A challenge for highly selective detection of them is that 5fU, 5fC, and AP all display respective formyl groups [21, 22]. These DNA formyl-modifications can all react with proteins via their active formyl groups to act as regulators in human's life [23–25]. Considering the low abundance of DNA formyl-modifications compared to canonical nucleobases in human genomic samples, even low levels of background noise are problematic [26]. Therefore, the highly selective labeling and fluorescence switch-on detections of 5fU and 5fC are in extremely urgent need. Here, we designed and synthesized a series of reagents for highly selective targeting and fluorescence switch-on detection of 5fU or 5fC [27–29].

2 Materials

1. All the oligonucleotides were synthesized and purified by GeneCreate Co., Ltd. (Wuhan, China) unless stated otherwise.
2. The oligonucleotides containing 5-formylcytosine were purchased from Takara Biotechnology (Dalian, China).
3. The oligonucleotides containing 5-formyluracil were synthesized using 5-formyluracil phosphoramidite (Self-synthesized) [30].
4. The oligonucleotides containing abasic sites were synthesized through ODN-U (one T site was replaced by uracil) treated with uracil DNA glycosylase (Invitrogen™, USA) [30].
5. DNA concentration was quantified by NanoDrop 2000c (Thermo Scientific, USA). DNA MALDI-TOF Mass Spectra were collected on MALDI-TOF-MS (Shimadzu, Japan).
6. Gel Imaging was collected in Pharos FX Molecular imager (Bio-Rad, USA).
7. The nucleic acid stains Super GelRed (NO.: S-2001) was purchased from US Everbright Inc. (Suzhou, China).
8. YeaRed Nucleic Acid Gel Stain (NO.: 10202ES76) was purchased from YEASEN Biotechnology Co. Ltd., (Shanghai, China).
9. UV absorption spectra were acquired with SHIMADZU UV-2550.
10. Fluorescence emission spectra were recorded on PerkinElmer LS 55 (PerkinElmer, USA).
11. pH was measured with Mettler Toledo, FE20-Five Easy™ pH (Mettler Toledo, Switzerland).
12. LC-MS data were collected with the Agilent™ 1220 Infinity LC combined with the 6120 Single Quadrupole mass spectrometer (Agilent Technologies).
13. Degradase Plus and enzyme reaction buffer were purchased from Zymo Research (Zymo Research, USA).
14. MES buffer (100 mM, pH 6.0) was purchased from Sigma-Aldrich.
15. PBS buffer was purchased from Sigma-Aldrich.
16. Polyethylene glycol 200 (PEG-200) was purchased from Sigma-Aldrich.
17. Thermo-shaker was purchased from Ningbo Biocotek Scientific Instrument Co., Ltd., China.
18. 2-Cyanomethylbenzimidazole (CAS Number: 4414-88-4) was purchased from Adamas-beta® (Shanghai, China)

19. Chemical regent 2-(5-chlorobenzo[d]thiazol-2-yl) acetonitrile (CBAN), standard 5fU, 5fC were self-synthesized [27].
20. Mini quick spin oligo columns (Roche) was purchased from Sigma-Aldrich.
21. Calf thymus DNA was purchased from Sigma-Aldrich.

3 Methods

3.1 Fluorescence Labeling Method for Detection of 5-Formylcytosine

We used the gamma ray to irradiate calf thymus DNA at different irradiation doses to obtain detection samples containing different amount of 5fC [27]. 5-Methylcytosine in calf thymus DNA can be irradiated via gamma ray such as ^{60}Co to yield 5fC [31]. And different irradiation doses can cause different amount of 5fC. We designed and synthesized 2-(5-chlorobenzo[d]thiazol-2-yl) acetonitrile (CBAN) as chemical regent to selectively react with 5fC because 5fC carries both 4′-amino and 5′-formyl groups, but other formyl-modified DNA do not have the same groups (Fig. 3). 5fC can undergo an expected cyclization with cyano group in reagent CBAN to give a fluorescent nucleobase, but others have the similar formyl groups cannot.

1. Calf thymus DNA (Sigma-Aldrich) was dissolved in ddH_2O to make the final concentration of 0.5 mg mL^{-1}.
2. After that, it was bubbled with oxygen for 1 h under room temperature (*see* **Note 1**).
3. Then calf thymus DNA solution was irradiated by the ^{60}Co irradiation facility at various doses (0–240 Gy at 1.84 Gy min^{-1}) (*see* **Note 2**).
4. Next, 50 μg of irradiated calf thymus DNA (0, 60, 120, 180, 240 Gy) were added to the final MES buffer (100 mM, pH 6.0) containing 50 mM CBAN (*see* **Note 3**) at 60 °C for 10 h in a thermo-shaker (850 rpm), respectively (*see* **Notes 3** and **4**).
5. Finally, the mixture was directly detected to record the fluorescence spectra via fluorometer using excitation wavelength at 389 nm and emission at 430 nm, respectively. The fluorescence spectra data can be calculated to 5fC quantification value by comparing with fluorescence spectra data of standard 5fC samples in the same conditions. Meanwhile, this result was also confirmed by using previously reported LC-MS/MS method [32]. 1.30 5fC moieties per 10^6 nucleotides per Gy (1.84 Gy min^{-1}, ^{60}Co irradiation) was obtained by our fluorescence labeling method compared with 1.27 5fC moieties per 10^6 nucleotides per Gy via LC-MS/MS method [27].

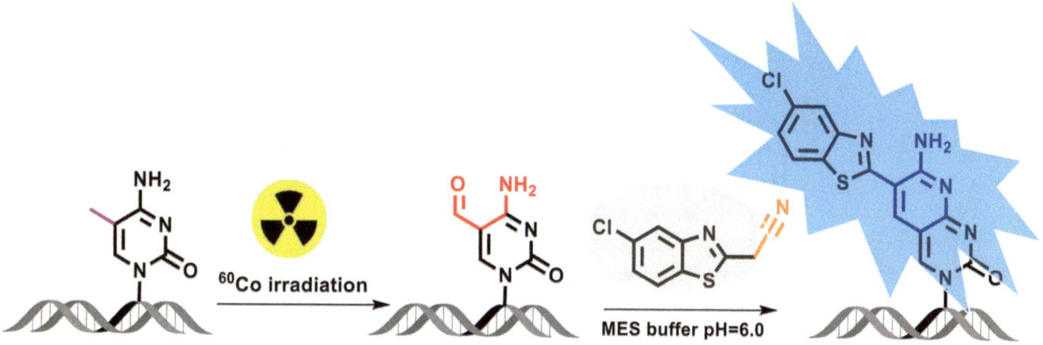

Fig. 3 5-Methylcytosine (5mC) is irradiated by ^{60}Co to yield 5-formylcytosine (5fC). 2-(5-chlorobenzo[d]thiazol-2-yl) acetonitrile (CBAN) can selectively react with 5fC via intramolecular cyclization with cyano reagents in CBAN and 4′-amino and 5′-formyl groups in 5fC to yield a fluorescent nucleobase

3.2 Fluorescence Labeling Method for Detection of 5-Formyluracil

We used synthesized DNA mixture which contained different amount of 5-formyluracil as detection samples. And we chose (2-benzimidazolyl)acetonitrile as chemical reagent for select tagging 5fU. (2-benzimidazolyl)acetonitrile could efficiently label 5fU with great selectivity and high yield under warm conditions in contrast to other formyl-modifications present in DNA (Fig. 4). The inherent chemical properties of (2-benzimidazolyl) acetonitrile make it highly selective for 5fU to yield a nucleobase which having aggregation-induced emission properties [9, 29].

1. (2-benzimidazolyl)acetonitrile was dissolved in DMSO to make the final concentration of 100 mM.
2. ODNs containing 5fU were performed in 100 mM NaOAc buffer (pH 5.0) with 12.5 mM (2-benzimidazolyl)acetonitrile at 37 °C for 6 h in a 1.5 mL tube in a thermo-shaker (Ningbo Biocotek Scientific Instrument Co., Ltd., China, 250 × g).
3. Mini quick spin oligo columns (Roche) were resuspended at column matrix and briefly centrifuge for 10 s.
4. Remove the spin oligo column top cap and snap off bottom tip and place column into microcentrifuge tube (*see* **Note 5**).
5. Spin column at 1000 *g* for 1 min to pack the column and remove residual buffer.
6. Add 100 μL of PBS buffer into the spin column and centrifuge at 1000 *g* for 2 min to change the previous buffer to PBS buffer (*see* **Note 6**).
7. Carefully apply sample DNA reaction mixture to the center of column bed and centrifuge at 1000 *g* for 4 min (*see* **Notes 7** and **8**).
8. Get the eluate solution and add PEG-200 to make a final mixture of PEG-200/PBS (v:v, 99:1). And vortex DNA mixture carefully for 30 s (*see* **Note 9**).

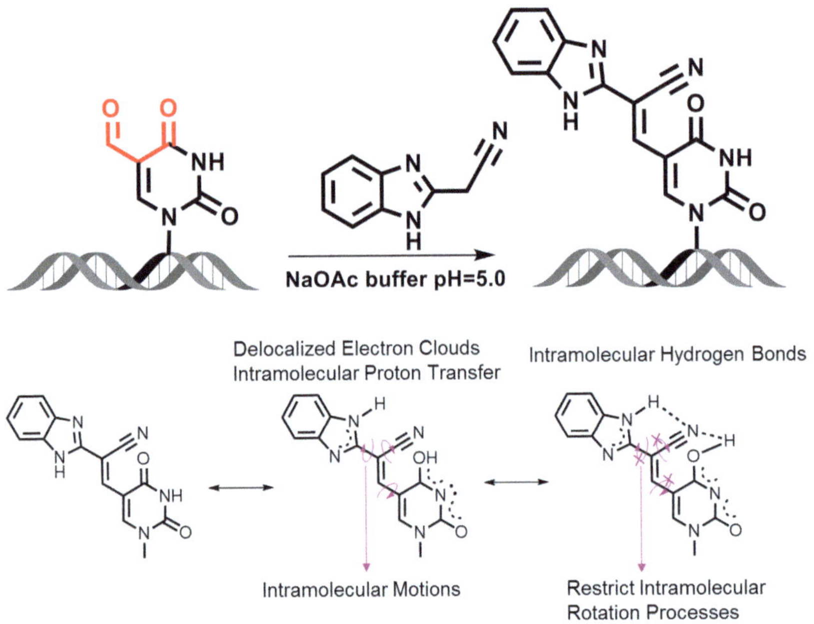

Fig. 4 (2-benzimidazolyl)acetonitrile can selectively tag 5-formyluracil (5fU) to yield a nucleobase which has aggregation-induced emission properties with high yield and under warm conditions. The yield nucleobase bears the non-radiative decay and characteristic of no fluorescence due to the molecular motions. However, the electron clouds are delocalized in the whole nucleobase. And the tunable intramolecular proton transfer can form intramolecular hydrogen bonds in aggregation status. Thus, intramolecular proton transfers together with the intramolecular hydrogen bonds served as noncovalent conformational locks to restrict intramolecular rotation processes, resulting a strong fluorescence

9. The DNA mixture was directly measured to record the fluorescence spectra via fluorimeter using excitation wavelength at 375 nm and emission at 430 nm, respectively. The fluorescence spectra data can be calculated to 5fU quantification value by comparing with fluorescence spectra data of standard 5fU samples in the same conditions. Correlation of the fluorescence intensity (at 430 nm) of 5fU after treatment with (2-benzimidazolyl)acetonitrile with various DNA concentrations has been calculated and the limit of detection is calculated as 157 nM [29].

4 Notes

1. While bubbling with oxygen into calf thymus DNA solution, pay attention to air pressure to avoid liquid splashing.
2. Subtle fluctuations in temperature while irradiation treatment of samples will influence the final amount of 5fC.
3. Due to the poor solubility of CBAN in the reaction buffer, it must be shaken sufficiently during the reaction.

4. The longer incubation time is not recommended since longer time will cause DNA damage. If the assay is only for qualitative detection of 5fC, lower temperature and shorter time may also work effectively. When optimize the temperature and incubation time, lower reaction temperature requires longer incubation time to ensure the whole 5fC to be converted.
5. If there are some bubbles in the Mini quick spin oligo columns before snapping off bottom tip, flick the column by hand carefully and then briefly centrifuge for 10 s to remove the bubbles.
6. Remove all elution buffer or change collection tube before adding DNA samples in Mini quick spin oligo columns.
7. Carefully apply sample DNA reaction mixture to the center of column bed to avoid the solution remaining on the pipette tips or the edge of the column.
8. If the volume exceeds 50 μL, use more than one column for changing buffer and keep all samples volume consistent.
9. Vortex DNA in PEG-200/PBS (v:v, 99:1) solution is necessary to ensure yielding nucleobase to perform the aggregation-induced emission properties before recording the fluorescence spectra.

References

1. Wang Y, Zhang X, Zou G, Peng S, Liu C et al (2019) Detection and application of 5-formylcytosine and 5-formyluracil in DNA. Acc Chem Res 52(4):1016–1024
2. Yuan B-F (2020) Assessment of DNA epigenetic modifications. Chem Res Toxicol 33 (3):695–708
3. Berney M, McGouran JF (2018) Methods for detection of cytosine and thymine modifications in DNA. Nat Rev Chem 2(11):332–348
4. Pais JE, Dai N, Tamanaha E, Vaisvila R, Fomenkov AI et al (2015) Biochemical characterization of a Naegleria TET-like oxygenase and its application in single molecule sequencing of 5-methylcytosine. Pro Natl Acad Sci U S A 112(14):4316–4321
5. Hong H, Wang Y (2007) Derivatization with Girard reagent t combined with LC-MS/MS for the sensitive detection of 5-formyl-2′-deoxyuridine in cellular DNA. Anal Chem 79 (1):322–326
6. Park Y, Peoples AR, Madugundu GS, Sanche L, Wagner JR (2013) Side-by-side comparison of DNA damage induced by low-energy electrons and high-energy photons with solid TpTpT trinucleotide. J Phys Chem B 117 (35):10122–10131
7. Dietzsch J, Feineis D, Höbartner C (2018) Chemoselective labeling and site-specific mapping of 5-formylcytosine as a cellular nucleic acid modification. FEBS Lett 592 (12):2032–2047
8. Jiang H-P, Liu T, Guo N, Yu L, Yuan B-F et al (2017) Determination of formylated DNA and RNA by chemical labeling combined with mass spectrometry analysis. Anal Chim Acta 981:1–10
9. Wang YF, Liu CX, Wu F, Zhang X, Liu S et al (2018) Highly selective 5-formyluracil labeling and genome-wide mapping using (2-Benzimidazolyl)acetonitrile probe. iScience 9:423–432
10. Zeng H, He B, Yi C (2019) Compilation of modern technologies to map genome-wide cytosine modifications in DNA. Chembiochem 20(15):1898–1905
11. Ito S, Shen L, Dai Q, Wu SC, Collins LB et al (2011) Tet proteins can convert 5-methylcytosine to 5-formylcytosine and 5-carboxylcytosine. Science 333(6047):1300–1303
12. Bachman M, Uribe-Lewis S, Yang X, Burgess HE, Iurlaro M et al (2015) 5-formylcytosine can be a stable DNA modification in mammals. Nat Chem Biol 11(8):555–U540
13. Booth MJ, Marsico G, Bachman M, Beraldi D, Balasubramanian S (2014) Quantitative

sequencing of 5-formylcytosine in DNA at single-base resolution. Nat Chem 6(5):435–440
14. Dai Q, Sanstead PJ, Peng CS, Han D, He C et al (2016) Weakened N3 hydrogen bonding by 5-formylcytosine and 5-carboxylcytosine reduces their base-pairing stability. ACS Chem Biol 11(2):470–477
15. Song C-X, Szulwach KE, Dai Q, Fu Y, Mao S-Q et al (2013) Genome-wide profiling of 5-formylcytosine reveals its roles in epigenetic priming. Cell 153(3):678–691
16. Xia B, Han D, Lu X, Sun Z, Zhou A et al (2015) Bisulfite-free, base-resolution analysis of 5-formylcytosine at the genome scale. Nat Methods 12(11):1047–1050
17. Rogstad DK, Heo J, Vaidehi N, Goddard WA, Burdzy A et al (2004) 5-formyluracil-induced perturbations of DNA function. Biochemistry 43(19):5688–5697
18. Wang YF, Liu CX, Yang W, Zou GR, Zhang X et al (2018) Naphthalimide derivatives as multifunctional molecules for detecting 5-formylpyrimidine by both PAGE analysis and dot-blot assays. Chem Commun 54(12):1497–1500
19. Zhou Q, Li K, Liu Y-H, Li L-L, Yu K-K et al (2018) Fluorescent Wittig reagent as a novel ratiometric probe for the quantification of 5-formyluracil and its application in cell imaging. Chem Commun 54(97):13722–13725
20. Zou GR, Liu CX, Cong C, Fang ZT, Yang W et al (2018) 5-Formyluracil as a cornerstone for aluminum detection in vitro and in vivo: a more natural and sustainable strategy. Chem Commun 54(93):13107–13110
21. Heck C, Michaeli Y, Bald I, Ebenstein Y (2019) Analytical epigenetics: single-molecule optical detection of DNA and histone modifications. Curr Opin Biotech 55:151–158
22. Ivancova I, Leone D-L, Hocek M (2019) Reactive modifications of DNA nucleobases for labelling, bioconjugations, and cross-linking. Curr Opin Chem Biol 52:136–144
23. Xu W, Boyd RM, Tree MO, Samkari F, Zhao L (2019) Mitochondrial transcription factor A promotes DNA strand cleavage at abasic sites. Pro Natl Acad Sci U S A 116 (36):17792–17799
24. Li F, Zhang Y, Bai J, Greenberg MM, Xi Z et al (2017) 5-formylcytosine yields DNA–protein cross-links in nucleosome core particles. J Am Chem Soc 139(31):10617–10620
25. Zou G, Liu C, Zeng W, Yang W, Zhang K et al (2020) Regulable DNA–protein interactions in vitro and vivo at epigenetic DNA Marks. CCS Chem 2(2):54–63
26. Liu CX, Luo XM, Chen YQ, Wu F, Yang W et al (2018) Selective labeling aldehydes in DNA. Anal Chem 90(24):14616–14621
27. Liu CX, Wang YF, Yang W, Wu F, Zeng W et al (2017) Fluorogenic labeling and single-base resolution analysis of 5-formylcytosine in DNA. Chem Sci 8(11):7443–7447
28. Liu CX, Wang YF, Zhang X, Wu F, Yang W et al (2017) Enrichment and fluorogenic labelling of 5-formyluracil in DNA. Chem Sci 8 (6):4505–4510
29. Liu CX, Zou GR, Peng S, Wang YF, Yang W et al (2018) 5-formyluracil as a multifunctional building block in biosensor designs. Angew Chem Int Ed 57(31):9689–9693
30. Liu CX, Chen YQ, Wang YF, Wu F, Zhang X et al (2017) A highly efficient fluorescence-based switch-on detection method of 5-formyluracil in DNA. Nano Res 10(7):2449–2458
31. Madugundu GS, Cadet J, Wagner JR (2014) Hydroxyl-radical-induced oxidation of 5-methylcytosine in isolated and cellular DNA. Nucleic Acids Res 42(11):7450–7460
32. Tang Y, Xiong J, Jiang H-P, Zheng S-J, Feng Y-Q et al (2014) Determination of oxidation products of 5-methylcytosine in plants by chemical derivatization coupled with liquid chromatography/tandem mass spectrometry analysis. Anal Chem 86(15):7764–7772

Part IV

Detection of Base J and 8-oxo-7,8-dihydroguanine

Chapter 15

Mass Spectrometry-Based Quantification of β-D-Glucosyl-5-Hydroxymethyluracil in Genomic DNA

Shuo Liu and Yinsheng Wang

Abstract

Base J, or β-D-glucosyl-5-hydroxymethyluracil, is the first hypermodified nucleobase that has been discovered in eukaryotes. It has been found in the nuclear DNA of several kinetoplastid parasites. Within these organisms, base J is formed in two putative steps, which include the initial recognition and hydroxylation of a specific thymine by J-binding proteins (JBP1 and JBP2) and the subsequent glycosylation by a glucosyltransferase (GT). Since this hyper-modification event only occurs in the genome of kinetoplastids but none of other higher eukaryotes, base J becomes a potential target for parasite-specific chemotherapy. However, the biological function of base J and its protein partners involved in J synthesis remains largely elusive. The need for a better understanding of the roles of base J in biological processing necessitates an accurate measurement of the physiological level of base J. Here we introduced the development of a surrogate internal standard and a reversed-phase HPLC coupled to electrospray ionization tandem mass spectrometry method for accurate measurement of base J and its precursor in trypanosomatid DNA. This novel application of reversed-phase HPLC-tandem mass spectrometry built a solid foundation for further exploring the molecular mechanism of base J biosynthesis and the roles of base J in biological processes.

Keywords Kinetoplastid, Base J, Hydroxymethyluracil, Tandem mass spectrometry

1 Introduction

Kinetoplastid flagellates contain an unusual DNA base J (β-D-glucosyl-5-hydroxymethyluracil) which was identified as the first hypermodified base in eukaryotic DNA [1]. This unique hypermodification occurs in unicellular kinetoplastids, including *Trypanosoma*, *Leishmania* species [2], and *Euglena gracilis* [3]. Within these organisms, base J replaces a specific portion of thymine in the genomic DNA. Other than unicellular protozoa, base J is undetectable in animals, plants, fungi, and a range of other simple eukaryotes [2]. Base J has been found primarily enriched in telomeric repeat regions [4–6] and, at a much lower frequency, in other forms of repetitive DNA sequences [7] as well as sequences between transcription units, namely internal J [8, 9]. In *Trypanosoma brucei*,

Bi-Feng Yuan (ed.), *DNA Modification Detection Methods*, Springer Protocols Handbooks,
https://doi.org/10.1007/978-1-0716-1229-3_15,

it is also localized in the expression sites of sub-telomeric variant surface glycoprotein (VSG) gene that produces a VSG coat to help *T. brucei* evade host immune response [10]. The unique occurrence of base J in the kinetoplastids rather than other higher eukaryotes renders base J a possible target of chemotherapy in parasites [11].

Base J is synthesized in a two-step pathway. In the first step, a thymidine residue is de novo hydroxylated by a thymidine hydroxylase (TH) [11]. The intermediate, 5-hydroxymethyl-2′-deoxyuridine (5-HmdU), is then glycosylated to form β-D-glucosyl-5-hydroxymethyl-2′-deoxyuridine (dJ) by a glucosyltransferase [12, 13]. J-binding proteins 1 and 2 (JBP1 and JBP2) are the two enzymes discovered in trypanosomes that are responsible for thymidine hydroxylation [9]. Although they both contain an N-terminal TH domain [14], only the unique J-binding domain on JBP1 allows it to recognize base J in DNA [15]. JBP2 contains an ATPase/DNA helicase-related SWI2/SNF2 domain that may be crucial for its activity [16]. Scholars recently have correlated the JBP proteins with the family of Fe(II)- and 2-oxoglutarate-dependent dioxygenases which also includes mammalian ten-eleven translocation (TET) proteins [17]. TET proteins are a group of dioxygenases that catalyze the oxidation of 5-methylcytosine (5-mC) in mammalian DNA [18]. TET enzymes-mediated conversion of 5-mC represents a key step in active cytosine demethylation in mammals [18]. The TH domains of JBP proteins are the only homolog of TET proteins in eukaryotes [19].

While most studies have been focused on the biosynthesis of base J, little is known about its biological function until the recent discovery that base J acts as an epigenetic mark to regulate gene transcription in kinetoplastids. Different from other eukaryotes, transcription units in kinetoplastids are polycistronically transcribed by RNA polymerase II (RNAP II) [20]. Whole-genome profiling of trypanosomes uncovered enriched localization of internal J within polycistronic gene clusters [8]. Depletion of JBP1/2 in *Trypanosoma cruzi* resulted in loss of base J and a consequential loosen chromatin structure at transcription start sites, which eventually increases transcription initiation [21]. Additionally, it has also been shown that loss of internal J in *Leishmania* also lead to massive read through at RNAP II-regulated transcriptional termination regions [22]. These evidences suggest that base J may mediate global regulation of transcription through creating more repressive chromatin structure. Furthermore, base J also principally represents telomeric repeats in kinetoplastids. However, whether base J has a telomeric function remains elusive [11].

To further explore the biological functions of base J and to verify its occurrence in other species, a highly sensitive and accurate detection method is required. A variety of analytical techniques have been employed so far to determine the abundance of base J, which include two-dimensional thin-layer chromatography [3], immunoblotting and Southern blotting [6], immunoprecipitation experiments [10], and ^{32}P-postlabeling [23]. Nevertheless, it is non-negligible that these methods are limited by their poor sensitivity and/or accuracy as well as their failure to provide structural information. For instance, van Leeuwen et al. [23] found ^{32}P-postlabeling only recovered 50% of the total J as a consequence of incomplete digestion of this bulky DNA modification during sample preparation. Although detention sensitivity could be enhanced by applying rabbit polyclonal antisera against base J [2], antibody production becomes another yield-limiting factor as the generation of more antisera is often tough [11]. To overcome these barriers in DNA modification research, mass spectrometry techniques have been introduced as one of the most powerful methods for quantitative analysis of DNA modifications [24]. For example, an HPLC coupled to tandem mass spectrometry (LC-MS/MS) method together with the use of stable isotope dilution was employed to dissect the functions of repair proteins in eliminating DNA damage from the mammalian genome [25]. Other than harmful DNA lesions, it has also been widely applied to quantify DNA epigenetic marks that are involved in epigenetic regulation of diverse biological processes and human diseases [26]. However, since there has been no report yet on the application of mass spectrometry for determining the endogenous levels of base J, we set out to develop a novel LC-MS/MS method with the use of a surrogate internal standard to accurately quantify base J in *T. brucei*. The intermediate product of base J biosynthesis, 5-HmdU, was quantified in parallel.

2 Materials

2.1 Solutions

1. 40 μM uridine diphosphoglucose (UDP-Glc): dissolve with H_2O.
2. T4-BGT reaction buffer: 50 mM potassium acetate, 20 mM Tris acetate, 10 mM magnesium acetate, 1 mM DTT, pH 7.9.
3. 50 mM triethylammonium acetate (TEAA): prepare an appropriate volume of the solution and adjust to pH 6.8 with acetic acid. Filter with a 0.2 μm filter before use.
4. 30% (v/v) acetonitrile in 50 mM TEAA: to prepare 1 L solution, add 300 mL of acetonitrile to 700 mL of 50 mM TEAA solution and mix well. Filter with a 0.2 μm filter before use.
5. 400 mM 1,1,1,3,3,3-hexafluoro-2-propanol (HFIP): dissolve 41.5 mL of HFIP in ~950 mL of water. While mixing vigorously adjust to pH 7.0 with triethylamine. Adjust volume to 1 L with water. Filter with a 0.2 μm filter before use.

6. Cell lysis buffer: 20 mM Tris, bring to pH 8.1 with HCl, 20 mM EDTA, 400 mM NaCl, and 1% SDS (1 g of SDS in 100 mL of solution).
7. Saturated NaCl solution: dissolve 25 g NaCl in 50 mL water. Vortex for a few seconds. Before use, spin at 1000 × *g* for 1 min and take the supernatant.
8. Chloroform/isoamyl alcohol (24:1, v/v).
9. 70% ethanol: take 70 mL of ethanol and add enough H_2O to bring the volume to 100 mL. Store at −20 °C.
10. TE buffer: 10 mM Tris, bring to pH 8.0 with HCl, 1 mM EDTA.
11. 1 μM [1,3-$^{15}N_2$–2′-D]-5-HmdU: previously synthesized [27].
12. 10 mM *erythro*-9-(2-hydroxy-3-nonyl)adenine (EHNA): dissolve 5 mg of EHNA hydrochloride in 1.59 mL of H_2O. Once prepared in solution, store aliquots in tightly sealed vials at −20 °C or below and used within 1 month.
13. Enzymatic digestion solution A: 300 mM sodium acetate and 10 mM zinc chloride; prepare the appropriate volume of the solution and adjust to pH 5.6 with acetic acid.
14. Enzymatic digestion solution B: 0.1 M Tris–HCl, bring to pH 8.9 with HCl.
15. 10 mM ammonium formate: dissolve 0.63 g of ammonium formate in 1 L of H_2O. Filter with a 0.2 μm filter before use.
16. 0.1% (v/v) formic acid in water: use glass pipet to transfer 1 mL of formic acid to 1 L of H_2O. Filter with a 0.2 μm filter before use.
17. 0.1% (v/v) formic acid in acetonitrile: use glass pipet to transfer 1 mL of formic acid to 1 L of acetonitrile. Filter with a 0.2 μm filter before use.
18. 2 mM ammonium formate in water: dissolve 0.126 g of ammonium formate in 1 L of H_2O. Filter with a 0.2 μm filter before use.
19. 2 mM ammonium formate in methanol: dissolve 0.126 g of ammonium formate in 1 L of methanol. Filter with a 0.2 μm filter before use.

2.2 DNA Oligomers and Enzymes

1. Oligodeoxyribonucleotides (ODNs, 5′-ATGGCGXGCTAT-3′).
   X_1 = 5-hydroxymethyl-2′-deoxycytidine (5-HmdC).
   X_2 = 5-hydroxymethyl-2′-deoxyuridine (5-HmdU).
2. T4 phage *β*-glucosyltransferase (T4-BGT).
3. Proteinase K (800 units/mL).

4. RNase A (10 mg/mL).
5. RNase T1 (25 units/μL).
6. Nuclease P1 (100 units/μL).
7. Alkaline phosphatase, Calf Intestinal (CIP).
8. Phosphodiesterase 1 and 2.

2.3 Columns

1. Zorbax SB-C18 column (0.5 mm × 250 mm, 5 μm in particle size, Agilent Technologies).
2. Aeris WIDEPORE-C18 column (4.6 × 250 mm, 3.6 μm beads, 200 Å pore size, Phenomenex).
3. Trapping column packed with Magic reversed-phase AQ-C18 material (150 μm × 40 mm, 5 μm beads, 120 Å pore size, Michrom BioResources).
4. Analytical column packed with Magic reversed-phase AQ-C18 material (75 μm × 200 mm, 5 μm beads, 300 Å pore size, Michrom BioResources).

2.4 Equipments

1. Agilent 1200 capillary HPLC.
2. Agilent EASY-nLC II system.
3. Thermo Scientific LTQ linear ion-trap mass spectrometer.
4. High-speed centrifuge.
5. Beckman HPLC system (pump module 125) with a UV detector (module 126).
6. Speed-vacuum concentrator.
7. Nanodrop or similar instruments to determine DNA concentration.
8. Centrifuge tubes: 1.5, 2.0, 15, and 50 mL.

2.5 Software

1. Windows-based Xcalibur software to process MS data.

3 Methods

3.1 Preparation of Surrogate Internal Standard

3.1.1 Synthesis and Purification

1. Mix the ODNs (1 μg) with UDP-Glc (40 μM) and T4-BGT (1 unit) in 50 μL of T4-BGT reaction buffer.
2. Incubate at 37 °C for 1 h followed by at 65 °C for 10 min.
3. Remove enzymes by adding 50 μL of chloroform.
4. Vortex for 30 s and spin at 10,000 × *g* for 5 min.
5. Gently transfer the top aqueous layer to a clean tube. Dry the sample by speed vacuum.
6. Prepare HPLC mobile phases.

   A: 50 mM TEAA (pH 6.8).

B: 30% (v/v) acetonitrile in 50 mM TEAA (pH 6.8).

7. Connect an Aeris WIDEPORE column to Agilent 1100 HPLC system with a UV detector monitoring at 260 nm.
8. Set the flow rate as 800 μL/min.
9. Set up a gradient from 0 to 25% B over 5 min and from 25 to 50% B over 40 min.
10. Reconstituted the desired ODNs in 50 μL of double distilled water and mix well.
11. Inject the sample into the HPLC system and start run.
12. Manually collect fractions that contain ODNs.
13. Dry the fractions by speed vacuum.

3.1.2 Verify the Identity of the Purified ODNs by LC-MS/MS

1. Prepare mobile phases for HPLC.
   A: 400 mM HFIP (pH 7.0).
   B: methanol.
2. Connect a Zorbax SB-C18 column to Agilent 1200 capillary HPLC coupled with an LTQ linear ion-trap mass spectrometer.
3. Set the flow rate as 8.0 μL/min.
4. Set up the elution gradient on HPLC: 0% B over 5 min, from 0 to 20% B over 5 min, and from 20 to 50% B over 40 min.
5. Set up the LTQ linear ion-trap mass spectrometer to acquire MS/MS for the fragmentation of the $[M-3H]^{3-}$ ions of ODNs harboring 5-gHmdC (*m/z* 1283.5) or 5-gHmdU (*m/z* 1283.8).
6. Set the parameters on the mass spectrometer as follows.
   Ion mode: negative-ion mode.
   Temperature of the ion transport tube: 300 °C.
   Sheath gas flow rate: 15 arbitrary units.
   Spray, capillary, and tube lens voltages: 4.0 kV, −10 V, and −100 V.
7. Dissolve the dried ODN fraction in 10 μL of double distilled water.
8. Load 5 μL of the sample onto the HPLC and meanwhile start run on an LTQ linear ion-trap mass spectrometer.
9. To confirm the presence of dJ or gHmdC-modified nucleotides in ODNs, assign the fragments ions yielded from the MS/MS of the $[M-3H]^{3-}$ parent ion especially at the seventh position counting from the 5′ terminus of the ODN (*see* **Note 1**).

3.2 DNA Extraction

1. Resuspend cell pellet (one million cells) in 200 μL of cell lysis buffer in a 1.5-mL tube.
2. Add proteinase K to a final concentration of 20 mg/mL.
3. Incubate the tube in water bath at 55 °C overnight.

4. Add ½ volume (100 μL) of saturated NaCl solution to the digestion mixture.
5. Vortex for 1 min and then incubated at 55 °C for 15 min.
6. Centrifuge at 10,000 × *g* for 15 min with a high-speed centrifuge.
7. Collect the supernatant and centrifuge again under the same condition as in **step 6**.
8. Precipitate the nucleic acids from the supernatant with ethanol.
9. Centrifuge at 10,000 × *g* for 5 min and air-dry the precipitate.
10. Redissolve the precipitate in 100 μL of double distilled water.
11. Add 2.5 μL of solution of RNase A (10 mg/mL) and 1 μL of solution of RNase T1 (25 units/μL) and incubate the solution at 37 °C for 1 h.
12. Extract with an equal volume (100 μL) of chloroform/isoamyl alcohol (24:1, v/v) twice.
13. Precipitate the nucleic acids from the aqueous layer by ethanol precipitation.
14. Centrifuge at 10,000 × *g* for 5 min.
15. Wash the pellet with 70% cold ethanol twice.
16. Dissolve the pellet in 100 μL of TE buffer.
17. Quantify the amount of genomic DNA by UV spectrophotometry.

3.3 Enzymatic Digestion of Genomic DNA

1. Add 100 fmol of the 5-gHmdC-containing ODN (surrogate standard for dJ quantification) to 1 μg of DNA, adjust the total volume to 90 μL with double distilled water (*see* **Notes 2–4**).
2. To the DNA-containing solution, add nuclease P1 (0.1 U/μg DNA), phosphodiesterase 2 (0.00025 U/μg DNA), and EHNA (1 mM in final concentration) that are premixed in 10 μL of enzymatic digestion solution A.
3. Incubate the digestion mixture at 37 °C for 48 h (*see* **Note 5**).
4. Add alkaline phosphatase (0.05 U/μg DNA) and phosphodiesterase 1 (0.0005 U/μg DNA) in 100 μL of enzymatic digestion solution B.
5. Incubate the digestion mixture at 37 °C for 2 h.
6. Add 2 pmol of [1,3-$^{15}N_2$–2′-D]-5-HmdU (the isotopically labeled standard for 5-HmdU quantification by LC-MS/MS/MS).
7. Remove proteins from the digestion mixture by extraction with an equal volume of chloroform.
8. Vortex for 30 s and spin at 10,000 × *g* for 5 min.
9. Gently transfer the top aqueous layer to a clean tube.

10. Dry the aqueous layer by speed-vacuum and store at −20 °C.

3.4 Off-Line HPLC Enrichment

1. Prepare 10 mM ammonium formate as HPLC mobile phase A.
2. Connect an Aeris WIDEPORE C18 column to a Beckman HPLC system.
3. Set up an isocratic elution gradient (100% A) at a flow rate of 800 μL/min.
4. Reconstitute dried nucleic acids in 50 μL of double distilled water and mix well.
5. Inject the sample into the HPLC system and start run.
6. Manually collect two fractions according to the HPLC trace depicted in Fig. 1: ~11 min for 5-HmdU and 28–30 min for dJ and 5-gHmdC (*see* **Note 6**).
7. Dry the fractions by speed vacuum (*see* **Note 7**).

3.5 Quantitative Analysis of Base J

3.5.1 LC-MS/MS Analysis of dJ

1. Prepare mobile phases for HPLC.
   A: 0.1% (v/v) formic acid in water.
   B: 0.1% (v/v) formic acid in acetonitrile.
2. Connect the trapping and analytical columns with an LTQ XL linear ion trap mass spectrometer coupled to Agilent EASY-nLC II system.
3. Set up an elution gradient of 40 min 0–10% B at a flow rate of 300 nL/min on analytical column.
4. Set up the mass spectrometer to acquire MS/MS for the fragmentation of the $[M+H]^+$ ions of 5-gHmdC (*m/z* 420) and dJ (*m/z* 421).

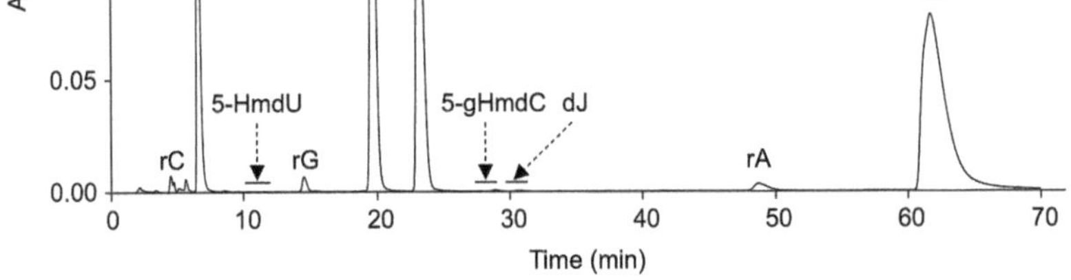

Fig. 1 Off-line HPLC trace for enrichment of 5-HmdU, 5-gHmdC, and dJ from *T. brucei* DNA. (Reproduced from ref. [28] with permission from American Chemical Society)

5. Set the parameters on LTQ XL linear ion trap mass spectrometer as follows.
   Ion mode: positive-ion mode.
   Temperature of the ion transport tube: 275 °C.
   Activation Q: 0.25.
   Activation time: 30 ms.
   Spray, capillary, and tube lens voltages: 2.0 kV, 12 V, and 100 V.
6. Reconstitute the fraction containing dJ and 5-gHmdC in 10 μL of double distilled water and well mix.
7. Load 4 μL of the sample onto the trapping column at a flow rate of 3 μL/min and elute directly to the analytical column. Meanwhile, start run on the LTQ XL linear ion trap mass spectrometer.
8. Extract selected-ion chromatogram from the raw MS file on Xcalibur by monitoring the unique transitions *m/z* 420 → 304 for 5-gHmdC and *m/z* 421 → 125, 143 for dJ. The proposed major fragmentation pathways for the $[M+H]^+$ ions of 5-gHmdC and dJ found in MS/MS are depicted in Fig. 2.
9. Calculate the peak area ratios of dJ versus 5-gHmdC for each measurement.

3.5.2 Calibration Curve Construction for dJ Quantification

1. Mix 0.1, 0.5, 1.5, 3, 6, 12.5, 25 pmol of the dJ-containing ODNs with 100 fmol of 5-gHmdC-harboring ODN and 1 μg of calf thymus DNA (*see* **Note 8**).
2. Repeat **step 1** on three separate days to obtain triplicate measurements for each sample.
3. Digest the samples as described in Subheading 3.3.
4. Enrich and isolate the ODN fractions containing dJ and gHmdC via offline HPLC following the procedure of Subheading 3.4.
5. Analyze 5-gHmdC- and dJ-containing HPLC fractions under the same LC-MS/MS conditions shown in Subheading 3.5.1.
6. Construct a calibration curve between the average ratio of peak areas (y) found in the selected-ion chromatograms for the analytes over their internal standards and the molar ratio (x) of the analytes over the standards added at **step 1**.
7. Derive an equation of the format $y = mx + b$ from the calibration curve, where m is the slope and b is the y-intercept. A representative calibration curve is depicted in Fig. 3.
8. Obtain a correlation coefficient (R^2) for the calibration curve. If $R^2 > 0.98$, apply the measurement of peak area ratio of dJ and 5-gHmdC in the digested sample to the equation and

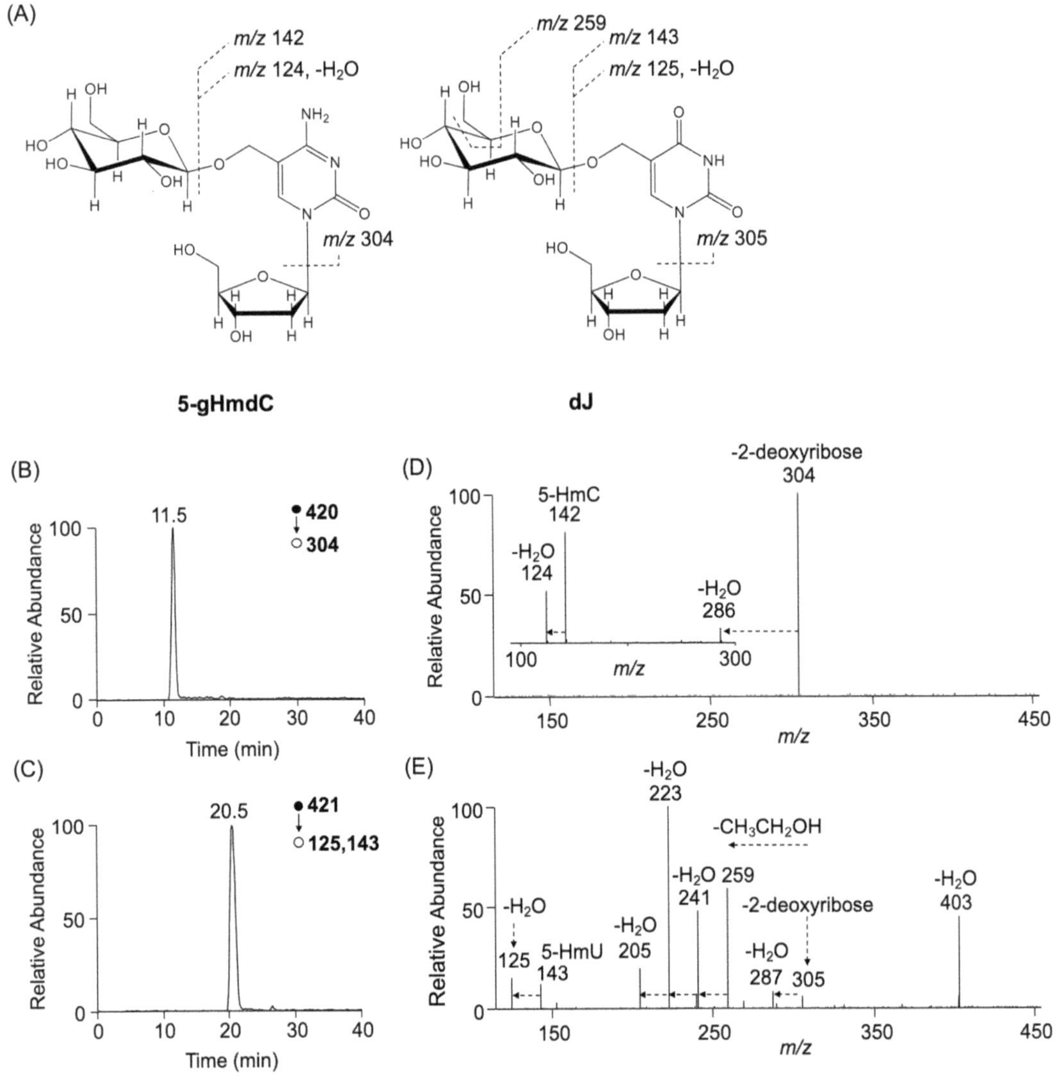

Fig. 2 LC-MS/MS for dJ quantification. (**a**) Chemical structures of dJ and 5-gHmdC and the cleavages for the formation of fragment ions observed in the MS/MS of the $[M+H]^+$ ion of 5-gHmdC and dJ. (**b** and **c**) Selected-ion chromatograms of the surrogate standard (5-gHmdC, **b**) and the analyte (dJ, **c**) detected in *T. brucei* genomic DNA. (**d** and **e**) MS/MS for 5-gHmdC (**d**) and dJ (**e**), and the inset of (**d**) gives MS/MS/MS of 5-gHmdC. (Reproduced from ref. [28] with permission from American Chemical Society)

calculate the number of moles of dJ and 5-HmdU in the sample.

9. Process the final data by dividing the number of moles of the modified nucleosides with the number of moles of total nucleosides in the digested DNA (*see* **Note 9**) and express the data in terms of numbers of modifications per 10^6 nucleosides.

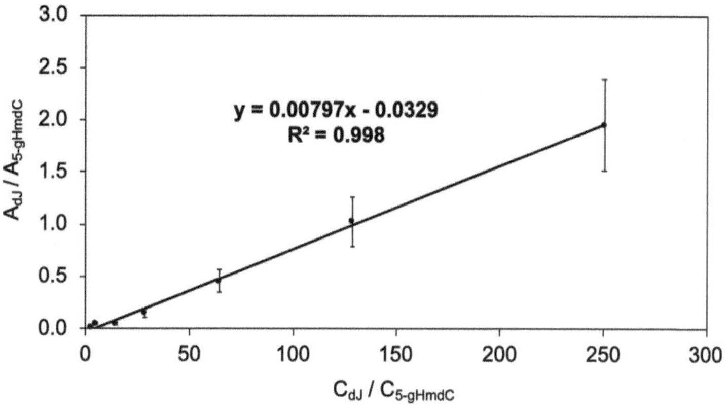

Fig. 3 The calibration curve for dJ quantification. Plotted are the ratios of peak areas found in the selected-ion chromatograms for monitoring the transitions as shown in Fig. 2b and c versus the molar ratios of dJ over that of 5-gHmdC standard (100 fmol for 5-gHmdC-containing ODN was used). The data represent mean ± standard deviation for measurements done in three days. (Reproduced from ref. [28] with permission from American Chemical Society)

3.6 Quantitative Analysis of 5-HmdU

1. Prepare mobile phases for HPLC.
   A: 2 mM ammonium formate in water.
   B: 2 mM ammonium formate in methanol.
2. Connect a Zorbax SB-C18 column with Agilent 1200 capillary HPLC pump. The effluent from the LC column was directed to LTQ linear ion-trap mass spectrometer.
3. Set the flow rate at 8 μL/min on HPLC.
4. Set up an elution gradient of 5 min 0–20% B and 25 min 20–70% B.
5. Set up the mass spectrometer to acquire MS/MS for the fragmentation of the $[M-H]^-$ ions of 5-HmdU (*m/z* 257) and [1,3-$^{15}N_2$–2′-D]-5-HmdU (*m/z* 260) as well as MS/MS/MS for the fragmentation of their major product ions observed in MS/MS (i.e., *m/z* 214 and *m/z* 216, respectively).
6. Set the parameters on the mass spectrometer as follows.
   Ion mode: negative-ion mode.
   Temperature of the ion transport tube: 300 °C.
   Sheath gas flow rate: 15 arbitrary units.
   Activation Q: 0.25.
   Spray, capillary, and tube lens voltages: 4.5 kV, −10 V, and −100 V.
7. Re-dissolve the dried fraction containing 5-HmdU in 10 μL of double distilled water.
8. Load 7 μL of the sample onto the HPLC and meanwhile start run on LTQ linear ion-trap mass spectrometer.

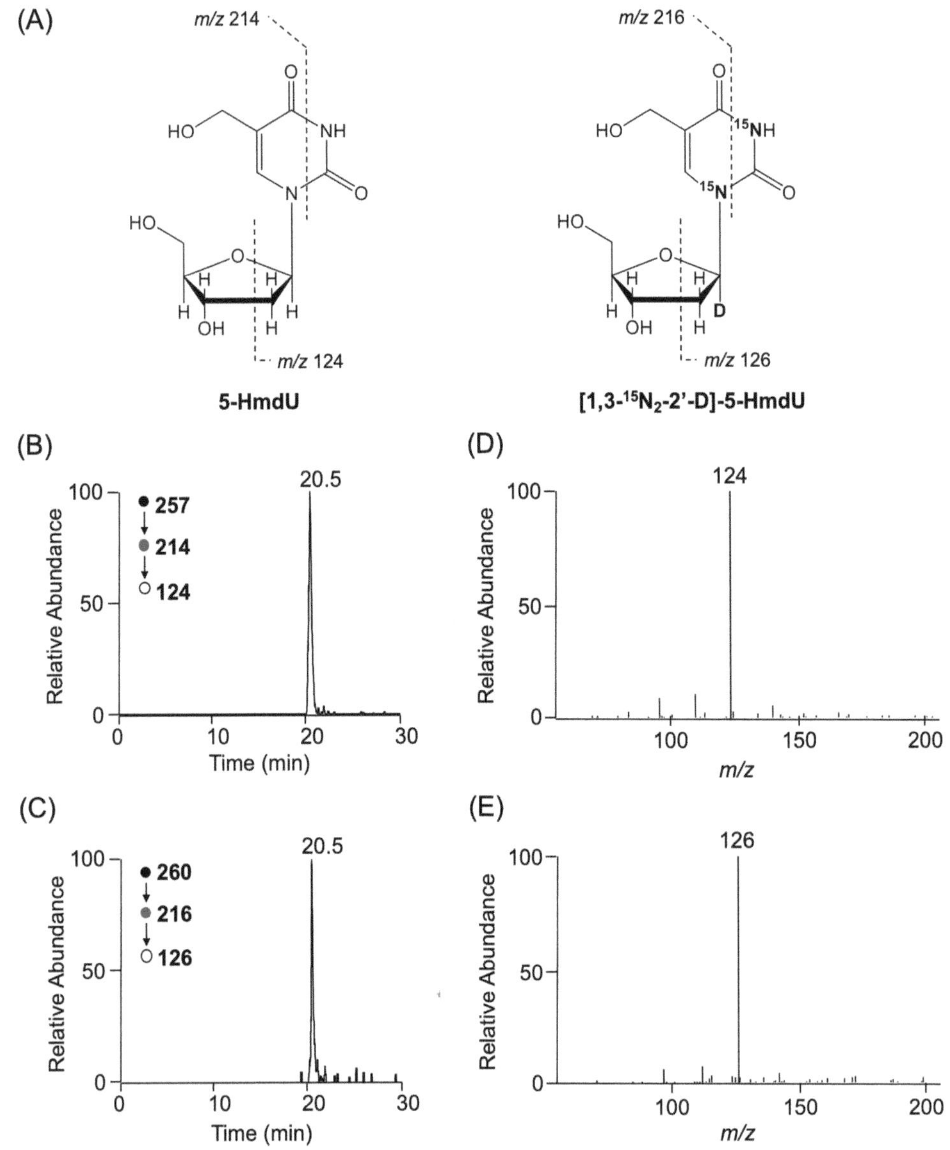

Fig. 4 LC-MS/MS/MS for 5-HmdU quantification. (**a**) Chemical structures and the cleavages for the formation of fragment ions observed in the MS/MS/MS of the $[M-H]^-$ ion of 5-HmdU. The ^{15}N and D isotopes are marked in the structure of [1,3-^{15} N_2–2′-D]-5-HmdU. (**b** and **c**) selected-ion chromatograms of unlabeled 5-HmdU (**b**) and isotope-labeled 5-HmdU (**c**) in *T. brucei* DNA. (**d** and **e**) MS/MS/MS for 5-HmdU (**d**) and isotope-labeled 5-HmdU (**e**). Reproduced from ref. [28] with permission from American Chemical Society

9. Extract selected-ion chromatogram from the raw MS file on Xcalibur by monitoring the unique transitions *m/z* 257 → 214 → 124 for 5-HmdU and *m/z* 260 → 216 → 126 for [1,3-^{15} N_2-2′-D]-5-HmdU. The proposed major fragmentation pathways for the $[M-H]^-$ ions of 5-HmdU found in MS/MS/MS are depicted in Fig. 4.

10. Calculate the peak area ratios of 5-HmdU over [1,3-^{15} N_2–2′-D]-5-HmdU for each sample.
11. Apply the peak area ratios to the calibration curve of 5-HmdU that was previously constructed [29] and calculate the number of moles of 5-HmdU in the samples.
12. Calculate the final data by dividing the number of moles of 5-HmdU with the number of moles of total nucleosides in the digested DNA and express the data in terms of numbers of modifications per 10^6 nucleosides.

4 Notes

1. The fragmentation of $[M-3H]^{3-}$ parent ion yields complementary fragments that are labeled as (a-base) and w ions [30]. In this study, fragmentation of the $[M-3H]^{3-}$ ions of 5-gHmdC-containing ODNs (*m/z* 1283.5) gives rise to a 481.5-Da difference between w_5 and ${w_6}^{2-}$ ions (*m/z* 1556.1 and *m/z* 1018.3, respectively) (Fig. 5). This mass difference is attributed to the presence of a 5-gHmdC monophosphate instead of a 2′-deoxycytidine monophosphate in the ODN, indicating its position at the seventh base from the 5′ terminus. The finding can also be confirmed by an observation of the mass difference between $[a_7\text{-}X]^{2-}$ and $[a_8\text{-}G]^{2-}$ ions (*m/z* 994.8 and *m/z* 1235.8, respectively) (Fig. 5).
2. Due to the difficulties in synthesizing isotope-labeled dJ in a multistep strategy, we decided to take an alternative approach to prepare the internal standard for quantifying endogenous dJ in nuclear DNA where a 5-gHmdC-harboring single-stranded ODN was designed and added as a surrogate standard.
3. The internal standard can only be used when 5-gHmdC is absent in the samples.
4. The 5-gHmdC-carrying ODN was digested with genomic DNA prior to off-line HPLC enrichment to avoid incomplete release of dJ from DNA during enzymatic hydrolysis and potential loss of the analyte during other stages of sample preparation, thereby providing more accurate quantification of dJ.
5. Seal the vials to minimize evaporation and check the volume for a few times to ensure that the solution isn't dried out during 48-h digestion.
6. An off-line HPLC enrichment was employed to filter unmodified nucleosides and buffer salts from the enzymatic digestion, which improves detection sensitivity and accuracy for dJ and 5-HmdU.

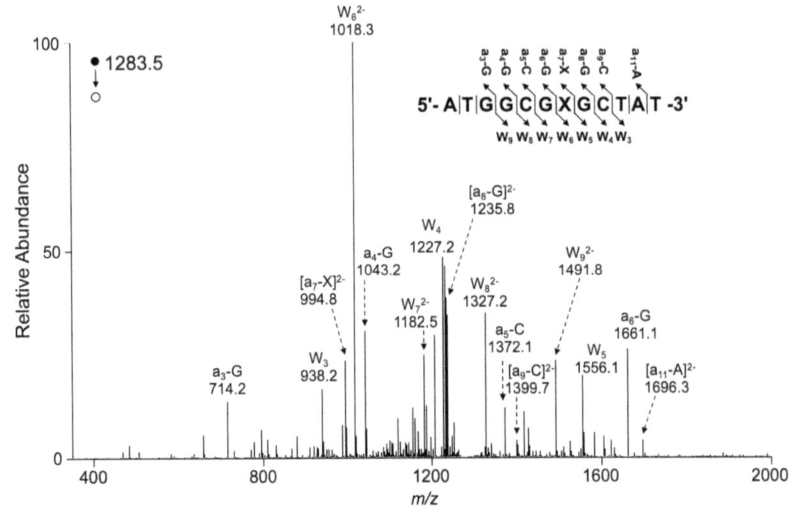

Fig. 5 Formation of fragment ions observed in the MS/MS of the $[M-3H]^{3-}$ ion of the standard ODN (5′-ATGGCGXGCTAT-3′, X = 5-gHmdC). (Reproduced from ref. [28] with permission from American Chemical Society)

7. If the fraction was collected and dried in more than one vials, combine all before subjecting to the mass spectrometer for each measurement.
8. Calf thymus DNA was added to the calibration mixture to mimic digestion conditions adapted to genomic DNA samples.
9. Considering potential RNA contamination in DNA samples, the percentage of RNA in DNA has been carefully calculated based on the peak areas of the 2′-deoxyribonucleosides and ribonucleosides found in the chromatogram of off-line HPLC enrichment and subsequentially subtracted from the total amount of nucleic acids.

References

1. Gommers-Ampt JH, Van Leeuwen F, de Beer AL, Vliegenthart JF, Dizdaroglu M, Kowalak JA, Crain PF, Borst P (1993) Beta-D-glucosyl-hydroxymethyluracil: a novel modified base present in the DNA of the parasitic protozoan T. brucei. Cell 75(6):1129–1136
2. van Leeuwen F, Taylor MC, Mondragon A, Moreau H, Gibson W, Kieft R, Borst P (1998) Beta-D-glucosyl-hydroxymethyluracil is a conserved DNA modification in kinetoplastid protozoans and is abundant in their telomeres. Proc Natl Acad Sci U S A 95(5):2366–2371
3. Dooijes D, Chaves I, Kieft R, Dirks-Mulder A, Martin W, Borst P (2000) Base J originally found in kinetoplastida is also a minor constituent of nuclear DNA of Euglena gracilis. Nucleic Acids Res 28(16):3017–3021
4. van Leeuwen F, Wijsman ER, Kuyl-Yeheskiely-E, van der Marel GA, van Boom JH, Borst P (1996) The telomeric GGGTTA repeats of Trypanosoma brucei contain the hypermodified base J in both strands. Nucleic Acids Res 24(13):2476–2482
5. Genest PA, Ter Riet B, Cijsouw T, van Luenen HG, Borst P (2007) Telomeric localization of the modified DNA base J in the genome of the protozoan parasite Leishmania. Nucleic Acids Res 35(7):2116–2124
6. Ekanayake DK, Cipriano MJ, Sabatini R (2007) Telomeric co-localization of the modified base J and contingency genes in the

protozoan parasite Trypanosoma cruzi. Nucleic Acids Res 35(19):6367–6377

7. van Leeuwen F, Kieft R, Cross M, Borst P (2000) Tandemly repeated DNA is a target for the partial replacement of thymine by beta-D-glucosyl-hydroxymethyluracil in Trypanosoma brucei. Mol Biochem Parasitol 109 (2):133–145
8. Cliffe LJ, Siegel TN, Marshall M, Cross GA, Sabatini R (2010) Two thymidine hydroxylases differentially regulate the formation of glucosylated DNA at regions flanking polymerase II polycistronic transcription units throughout the genome of Trypanosoma brucei. Nucleic Acids Res 38(12):3923–3935
9. Cliffe LJ, Kieft R, Southern T, Birkeland SR, Marshall M, Sweeney K, Sabatini R (2009) JBP1 and JBP2 are two distinct thymidine hydroxylases involved in J biosynthesis in genomic DNA of African trypanosomes. Nucleic Acids Res 37(5):1452–1462
10. van Leeuwen F, Wijsman ER, Kieft R, van der Marel GA, van Boom JH, Borst P (1997) Localization of the modified base J in telomeric VSG gene expression sites of Trypanosoma brucei. Genes Dev 11(23):3232–3241
11. Borst P, Sabatini R (2008) Base J: discovery, biosynthesis, and possible functions. Annu Rev Microbiol 62:235–251
12. Bullard W, Lopes da Rosa-Spiegler J, Liu S, Wang Y, Sabatini R (2014) Identification of the glucosyltransferase that converts hydroxymethyluracil to base J in the trypanosomatid genome. J Biol Chem 289(29):20273–20282
13. Sekar A, Merritt C, Baugh L, Stuart K, Myler PJ (2014) Tb927.10.6900 encodes the glucosyltransferase involved in synthesis of base J in Trypanosoma brucei. Mol Biochem Parasitol 196(1):9–11
14. Yu Z, Genest PA, ter Riet B, Sweeney K, DiPaolo C, Kieft R, Christodoulou E, Perrakis A, Simmons JM, Hausinger RP, van Luenen HG, Rigden DJ, Sabatini R, Borst P (2007) The protein that binds to DNA base J in trypanosomatids has features of a thymidine hydroxylase. Nucleic Acids Res 35 (7):2107–2115
15. Cross M, Kieft R, Sabatini R, Wilm M, de Kort M, van der Marel GA, van Boom JH, van Leeuwen F, Borst P (1999) The modified base J is the target for a novel DNA-binding protein in kinetoplastid protozoans. EMBO J 18(22):6573–6581
16. Kieft R, Brand V, Ekanayake DK, Sweeney K, DiPaolo C, Reznikoff WS, Sabatini R (2007) JBP2, a SWI2/SNF2-like protein, regulates de novo telomeric DNA glycosylation in bloodstream form Trypanosoma brucei. Mol Biochem Parasitol 156(1):24–31
17. Tahiliani M, Koh KP, Shen Y, Pastor WA, Bandukwala H, Brudno Y, Agarwal S, Iyer LM, Liu DR, Aravind L, Rao A (2009) Conversion of 5-methylcytosine to 5-hydroxymethylcytosine in mammalian DNA by MLL partner TET1. Science 324 (5929):930–935
18. Ito S, D'Alessio AC, Taranova OV, Hong K, Sowers LC, Zhang Y (2010) Role of Tet proteins in 5mC to 5hmC conversion, ES-cell self-renewal and inner cell mass specification. Nature 466(7310):1129–1133
19. Wu SC, Zhang Y (2010) Active DNA demethylation: many roads lead to Rome. Nat Rev Mol Cell Biol 11(9):607–620
20. Ekanayake D, Sabatini R (2011) Epigenetic regulation of polymerase II transcription initiation in Trypanosoma cruzi: modulation of nucleosome abundance, histone modification, and polymerase occupancy by O-linked thymine DNA glucosylation. Eukaryot Cell 10 (11):1465–1472
21. Ekanayake DK, Minning T, Weatherly B, Gunasekera K, Nilsson D, Tarleton R, Ochsenreiter T, Sabatini R (2011) Epigenetic regulation of transcription and virulence in Trypanosoma cruzi by O-linked thymine glucosylation of DNA. Mol Cell Biol 31 (8):1690–1700
22. van Luenen HG, Farris C, Jan S, Genest PA, Tripathi P, Velds A, Kerkhoven RM, Nieuwland M, Haydock A, Ramasamy G, Vainio S, Heidebrecht T, Perrakis A, Pagie L, van Steensel B, Myler PJ, Borst P (2012) Glucosylated hydroxymethyluracil, DNA base j, prevents transcriptional readthrough in leishmania. Cell 150(5):909–921
23. van Leeuwen F, de Kort M, van der Marel GA, van Boom JH, Borst P (1998) The modified DNA base beta-D-glucosylhydroxymethyluracil confers resistance to micrococcal nuclease and is incompletely recovered by 32P-postlabeling. Anal Biochem 258 (2):223–229
24. Dudley E, Bond L (2014) Mass spectrometry analysis of nucleosides and nucleotides. Mass Spectrom Rev 33(4):302–331
25. Liu S, Wang Y (2013) A quantitative mass spectrometry-based approach for assessing the repair of 8-methoxypsoralen-induced DNA interstrand cross-links and monoadducts in mammalian cells. Anal Chem 85 (14):6732–6739
26. Yuan B-F, Feng Y-Q (2014) Recent advances in the analysis of 5-methylcytosine and its

oxidation products. TrAC-Trends Anal Chem 54(0):24–35

27. Hong H, Cao H, Wang Y, Wang Y (2006) Identification and quantification of a guanine-thymine intrastrand cross-link lesion induced by Cu(II)/H2O2/ascorbate. Chem Res Toxicol 19(5):614–621
28. Liu S, Ji D, Cliffe L, Sabatini R, Wang Y (2014) Quantitative mass spectrometry-based analysis of beta-D-glucosyl-5-hydroxymethyluracil in genomic DNA of Trypanosoma brucei. J Am Soc Mass Spectrom 25(10):1763–1770
29. Wang J, Yuan B, Guerrero C, Bahde R, Gupta S, Wang Y (2011) Quantification of oxidative DNA lesions in tissues of Long-Evans Cinnamon rats by capillary high-performance liquid chromatography-tandem mass spectrometry coupled with stable isotope-dilution method. Anal Chem 83 (6):2201–2209
30. McLuckey SA, Van Berkel GJ, Glish GL (1992) Tandem mass spectrometry of small, multiply charged oligonucleotides. J Am Soc Mass Spectrom 3(1):60–70

Chapter 16

Determination of 8-Oxo-7,8-Dihydroguanine in DNA at Single-Base Resolution by Polymerase-Mediated Differential Coding

Feng Tang and Bi-Feng Yuan

Abstract

Reactive oxygen species (ROS) can induce DNA damages in cells. 8-Oxo-7,8-dihydroguanine (OG) is viewed as one of the most frequent oxidative modifications in human genomes. It was reported that OG was also capable to facilitate the G-quadruplex formation and participate in the transcription process. Thus, OG might have potential functions in regulating gene expression. To investigate the molecular mechanisms of OG on affecting gene expression in vivo, it is necessary to determine the location of OG in DNA beforehand. Herein, we characterized Bsu DNA polymerase (Bsu Pol) and Tth DNA polymerase (Tth Pol), which can faithfully incorporate dATP or error-prone incorporate dCTP when bypassing OG, respectively. Based on the different coding performance, we achieved single-base resolution analysis of OG in DNA, which offers a promising approach for high-throughput analysis of OG at genome-wide scale.

Keywords 8-Oxo-7,8-dihydroguanine, DNA polymerase, Reactive oxygen species, Epigenetics, Sequencing, Single-base resolution

1 Introduction

Endogenous metabolism and exposure to various environmental pollutants and radiation can result in the elevated formation of reactive oxygen species (ROS) in human cells [1]. ROS can attack proteins, lipids, and nucleic acids [2]. Guanine has the lowest redox potential among four DNA bases, which makes guanine the most susceptible target of ROS [3, 4]. 8-Oxo-7,8-dihydroguanine (OG) is a kind of predominant oxidative lesion in DNA (Fig. 1a). Instead of blocking DNA replication and transcription, OG tends to induce G → T transversion mutations, resulting in cancers or other diseases [5].

DNA and RNA contain a variety of modifications that can control and regulate gene expression [7–15]. OG occurring in DNA has been traditionally viewed as mutagenic and detrimental to cellular processes [16]. Recent studies showed that OG may also

Bi-Feng Yuan (ed.), *DNA Modification Detection Methods*, Springer Protocols Handbooks,
https://doi.org/10.1007/978-1-0716-1229-3_16,

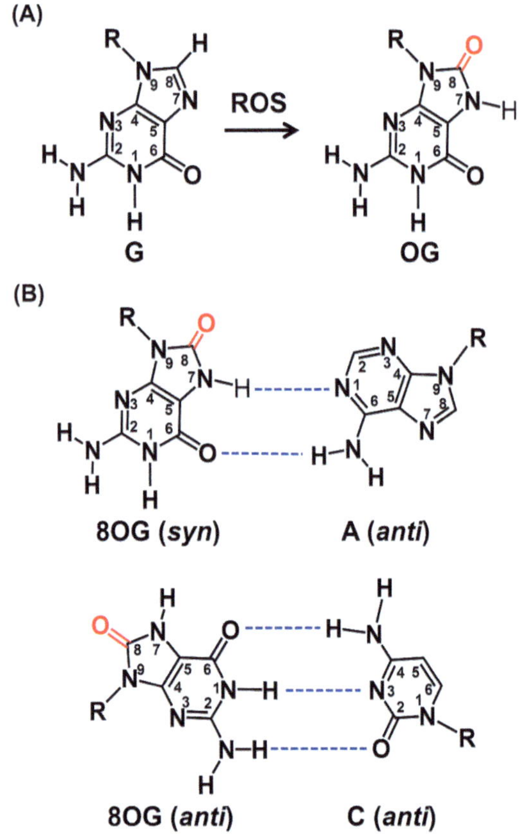

Fig. 1 The formation of OG from G by ROS and the base pairing of OG with A and C. (**a**) Oxidation of G at C8 to form OG induced by ROS. (**b**) OG (*syn*) in a Hoogsteen base pairing with A (anti); OG (anti) in a Watson–Crick base pairing with C (*anti*). Dashed lines indicate hydrogen bonds. (Reproduced from ref. [6] with permission from The Royal Society of Chemistry)

participate in modulating gene expression [17]. Pan et al. [18] reported that the OG formation in the genome can promote gene expression via the base excision repair (BER) pathway. The transcriptionally active euchromatin DNA harbored more OG than that in transcriptionally silenced heterochromatin in porcine thymus DNA [19]. Recently, it was found that OG can stimulate DNA transcription by facilitating the formation of G4 structure in the promoter region [20]. These studies indicate that OG may have regulatory and epigenetic-like properties [21].

OG can adopt two alternative conformations (*anti* or *syn*) in the active site of DNA polymerases, which endows OG the dual coding potential [22]. The *anti* conformation allows OG to pair with a cytosine via Watson–Crick base pair; on the contrast, the *syn* conformation of OG forms a stable pairing with an adenine in a common *anti* conformation by Hoogsteen base pair (Fig. 1b). After characterized some DNA polymerases, we found that *Bsu*

DNA polymerase (*Bsu* Pol) mainly incorporated dATP opposite OG and *Tth* DNA polymerase (*Tth* Pol) predominantly incorporated dCTP opposite OG [6]. We conducted an approach to localize OG in DNA at single-base resolution based on the distinct properties of these two DNA polymerases.

2 Materials

All solutions are prepared using Milli-Q water (Milli-Q apparatus, Millipore). All solvents and chemicals are of analytical grade.

2.1 Oligonucleotides

The 134-mer OG-containing single-stranded DNA (L-DNA-OG1, L-DNA-OG2, L-DNA-OG3) and the control 134-mer single-stranded DNA (L-DNA-G), were purchased from Takara Biotechnology Co., Ltd. (Dalian, China). The 51-mer DNA (NN-OG-NN, N represents random bases), with two randomized bases around the OG site was purchased from Integrated DNA Technologies (Iowa, USA). The other oligonucleotides were synthesized and purified by Sangon Biotech Co., Ltd. (Shanghai, China). DNA sequences are listed in Table 1.

2.2 Chemicals and Reagents

1. *Bsu* DNA polymerase (*Bsu* Pol, New England Biolabs, USA).
2. *Tth* DNA polymerase (*Tth* Pol, Toyobo Life Science Co., Ltd., China).
3. 2′-Deoxyribonucleoside triphosphates (dNTPs) (Takara Biotechnology Co., Ltd., China) (*see* **Note 1**).
4. Reaction buffer (10 mM Tris–HCl, 10 mM $MgCl_2$, 50 mM NaCl and 1 mM DTT, pH 7.9).
5. Exonuclease I (New England Biolabs, USA).
6. 2× TSINGKE Master mix (TSINGKE, China).
7. QIAquick PCR purification kit (QIAGEN).
8. E.Z.N.A.® Genomic DNA isolation kit (Omega Bio-Tek Inc., USA).

2.3 Equipment

1. Pharos FX Molecular imager (Bio-Rad, USA).
2. NanoDrop 2000c (Thermo Scientific, USA).

Table 1
The sequences of the oligodeoxynucleotides. (Reproduced from ref. [6] with permission from The Royal Society of Chemistry)

ODNs	Sequence (from 5′ to 3′)
FAM-primer-1	5′-FAM-CGCATAACCCTAACC-3′
DNA-G	5′-GCGTATTGGGATTGG**G**ATTGACACG-3′
DNA-OG	5′-GCGTATTGGGATTGG(**OG**)ATTGACACG-3′
Cy3-primer	5′-Cy3-TAACCCTAACCCTAACC-3′
Cy5-primer	5′-Cy5-CCTAACCCTAACCCTAAC-3′
FAM-primer-2	5′-FAM-CCCTAACCCTAACCCTAA-3′
T-DNA-1	5′-GTTA**GGG**TTAGGGTTAGGGTTAGGG-3′
T-DNA-2	5′-GTTA(**OG**)**GG**TTAGGGTTAGGGTTAGGG-3′
T-DNA-3	5′-GTTA**G**(**OG**)**G**TTAGGGTTAGGGTTAGGG-3′
T-DNA-4	5′-GTTA**GG**(**OG**)TTAGGGTTAGGGTTAGGG-3′
L-DNA-G	5′-*GCCCAAGTGCTGAGGCTGATAA*TAATCGG GGCGGCGATCAGAC AGCCCCGGTGTGGGAAA TCGTCCGCCCGGTCTCCCTAAGTCCCCGAAGTC GCCTCCCACTTTTGGT**G**ACT*GCTTGTTTA* *TTTACATGCAGinvertedT*-3′
L-DNA-OG	5′-*GCCCAAGTGCTGAGGCTGATAA*TAA TCGGGGCGG CGATCAGACAGCCCCGGTGTGGGAAATCGTC CGCCCGGTCTCCCT AAGTCCCCGAAGTCGCCTCCCA CTTTTGGT(**OG**)ACT*GCTTGTTTA* *TTTACATGCAGinvertedT*-3′

L-primer	5′-TGAGCAGATGTGTGACGGCTACA CTGCATGTAAATAAACAAGC-3′
PCR reverse primer	5′-GCCCAAGTGC TGAGGCTGATAA-3′
PCR forward primer	5′-TGAGCAGATGTGTGACGGCTAC-3′
NN-OG-NN	5′-*CCAGATGACGACTGGCAC*TA ATGNN(**OG**)NNTGCTG*CGTG GATCGACATCGAGinvertedC*-3′
NN-L-Primer	5′-TGAGCAGATGTGTGACGGCTACGC TCGATGTCGATCCACG-3′
NN-F-Primer	5′-CCAGATGACGACTGGCAC-3′
NN-R-Primer	5′-TGAGCAGATGT GTGACGGCTAC-3′

3 Methods

3.1 Quantitative Analysis of 8OG in DNA

3.1.1 Steady-State Kinetics Study

1. FAM-primer-1 (1.0 μM) was annealed with DNA-G or DNA-OG (1.0 μM) in reaction buffer (10 mM Tris–HCl, 50 mM NaCl, 10 mM $MgCl_2$ and 1 mM DTT, pH 7.9) by heating at 95 °C for 5 min and then cooling down slowly to 25 °C.
2. DNA polymerase (0.5 U) was added to the above mixture. The reaction was initiated by the addition of dATP/dCTP (2–4 mM) and the mixture (10 μL) was incubated at 37 °C for 1–20 min.
3. The reaction was quenched by adding 20 μL of stop buffer (25 mM EDTA, 95% formamide, pH 8.0), and the mixture was then heated for 10 min at 95 °C.
4. After cooling down to 25 °C, the mixture was analyzed by 20% denaturing polyacrylamide gel electrophoresis (PAGE, acrylamide/bisacrylamide = 19/1) in TBE buffer.
5. The bands were visualized and quantified by using the Pharos FX Molecular imager (Bio-Rad, USA).
6. The relative reaction velocity (v) was calculated from the following formula: $v \times t = I_E/(I_U + I_E)$, where I_E represents the extended product, I_U is the unextended primer, and t represents the reaction time.
7. The apparent K_M and V_{max} values were obtained according to previous described method [23]. By using the data points at different dATP concentrations in three independent experiments, the K_M and V_{max} were calculated based on the linear regression analysis of Hanes–Woolf plots.
8. V_{max}/K_M represents the selectivity of DNA polymerase for incorporating dATP/dCTP opposite OG or G (Table 2).

3.1.2 Quantitative Evaluation of OG in DNA via Primer Extension

1. Mix different ratios of DNA-OG/DNA-G with the percentage of DNA-OG ranging from 0% to 100%.
2. FAM-primer-1 (1.0 μM) and the template DNA (1.0 μM, 0–100% of OG) were dissolved in reaction buffer (10 mM Tris–HCl, 10 mM $MgCl_2$, 50 mM NaCl, and 1 mM DTT, pH 7.9) and annealed at 95 °C for 5 min and then cooled down to 25 °C.
3. Add *Bsu* DNA polymerase (0.5 U) to the mixture.
4. The reaction was initiated by adding the dATP (100 μM) to the mixture (10 μL). Incubate the mixture for 20 min at 37 °C (*see* **Note 2**).

Table 2
Steady-state kinetics for primer extension of the G- or OG-containing DNA template with different DNA polymerases. (Reproduced from ref. [6] with permission from The Royal Society of Chemistry)

DNA polymerase	dNTP	DNA template	V_{max} (% min^{-1})	K_M (μM)	V_{max}/K_M (min^{-1} μM^{-1})
Bsu	dATP	DNA-G	14.5 ± 2.3	178.6 ± 4.9	0.8×10^{-3}
	dATP	DNA-OG	25.1 ± 2.4	9.2 ± 0.6	27.3×10^{-3}
	dCTP	DNA-G	42.2 ± 0.6	1.8 ± 0.1	234.4×10^{-3}
	dCTP	DNA-OG	27.8 ± 2.3	107.5 ± 6.3	2.6×10^{-3}
Tth	dATP	DNA-G	5.4 ± 0.5	228.3 ± 15.4	0.2×10^{-3}
	dATP	DNA-OG	6.2 ± 0.6	275.8 ± 18.2	0.2×10^{-3}
	dCTP	DNA-G	50.4 ± 2.6	2.5 ± 0.3	201.6×10^{-3}
	dCTP	DNA-OG	29.7 ± 1.4	174.3 ± 23.5	1.7×10^{-3}

V_{max}: the maximum rate of the enzyme reaction
K_M: the Michaelis constant

5. The extension rate of FAM-primer-1 at each percentage of DNA-OG was calculated from the formula: $v \times t = I_E/(I_U + I_E)$.
6. Construct the calibration curve of OG by plotting the mean extension rate versus the mean molar ratio of OG/G (ranging from 0% to 100% of OG/G) based on data obtained from triplicate measurements (Fig. 2).

3.1.3 Analysis of 8OG in Telomeric DNA from HeLa Cells

1. HeLa cells were obtained from the China Center for Type Culture Collection (CCTCC) and maintained in Dulbecco's Modified Eagle medium (DMEM) supplemented with 10% FBS, 100 U/mL penicillin, and 100 μg/mL streptomycin (GIBCO) under 5% CO_2 atmosphere at 37 °C.
2. The cells were grown to ~80% confluence in 75-cm^2 culture flasks, washed with phosphate-buffered saline (PBS), and then treated with medium including 1000 μM H_2O_2. After incubation for 30 min in a CO_2 incubator, cells were harvested follow by washing twice.
3. Genomic DNA was isolated using the E.Z.N.A.® Genomic DNA isolation kit (Omega Bio-Tek Inc.) following the manufacturer's instruction. Deferoxamine mesylate was added in all the solutions as the oxidation inhibitor for protecting DNA from excessive oxidation during the extraction (*see* **Note 3**).
4. Determine the concentrations of the purified DNA by NanoDrop 2000c (Thermo Scientific, USA).

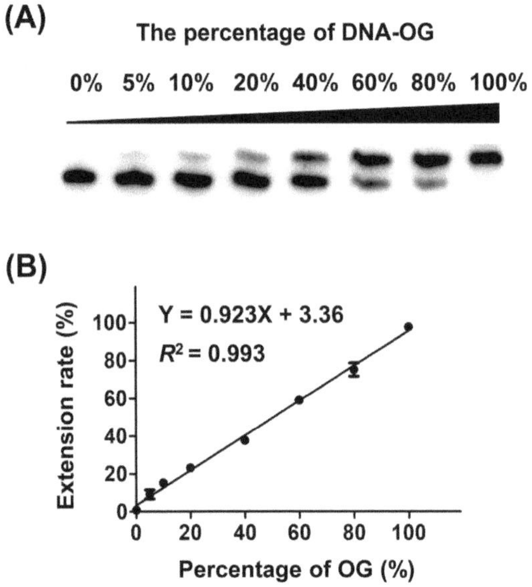

Fig. 2 Quantification of the OG in DNA. (**a**) Single-nucleotide primer extension reaction using the mixture of DNA-OG and DNA-G with the percentage of DNA-OG ranging from 0% to 100%. (**b**) The extension rate of FAM-primer-1 is proportional to the relative amounts of OG. Error bars are standard deviations from three independent experiments. (Reproduced from ref. [6] with permission from The Royal Society of Chemistry)

5. To measure the content of OG in the telomeric DNA, single-nucleotide primer extension assay was performed with the mixture of HeLa DNA (5 μg) and different fluorophore-labeled primers (Cy3-primer, Cy5-primer, and FAM-primer-2,5 pmol for each) (Fig. 3a).
6. The percentage of the extension of primers was calculated from the following formula: $I_E/(I_U + I_E)$ (Fig. 3b).

3.2 Analysis of OG by Sequencing

3.2.1 Analysis of OG in Synthesized DNA by Sanger Sequencing

1. The mixture of L-DNA-G (100 nM) or L-DNA-OG (100 nM) and L-primer (100 nM) in reaction buffer (10 mM Tris–HCl, 10 mM $MgCl_2$, 50 mM NaCl, and 1 mM DTT, pH 7.9) was incubated for 5 min at 95 °C and then cool down to 25 °C.
2. Add dNTPs (the final concentration is 100 μM for each) and *Bsu/Tth* polymerase (0.5 U). The mixture was incubated for 20 min at 37 °C.
3. The extension product was treated with exonuclease I (15 U) for 15 min at 37 °C to eliminate the L-primer (*see* **Note 4**).
4. The mixture was purified by QIAquick PCR purification kit (QIAGEN) by following the manufacturer's instruction.
5. The purified product was used as template for PCR amplification. PCR was performed by using 2× TSINGKE Master mix (TSINGKE, Beijing, China) according to the manufactures' instruction.

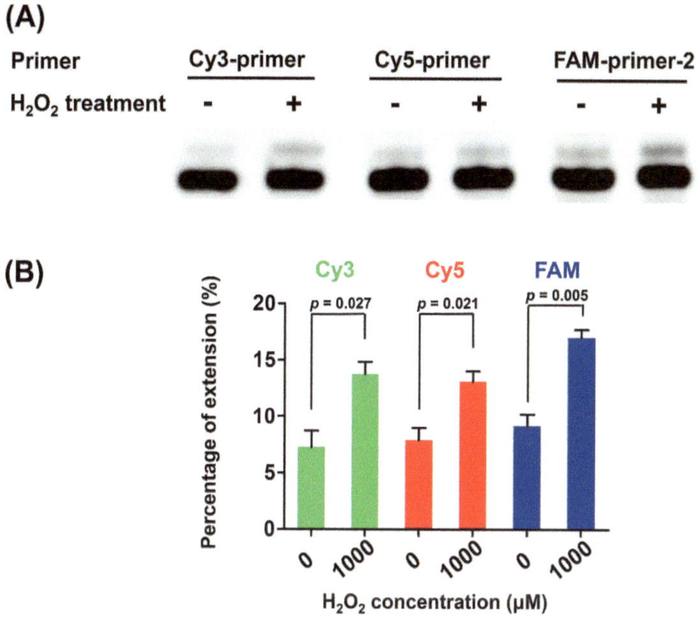

Fig. 3 Measurement of the level of OG in telomeric DNA. (**a**) The single-nucleotide primer extension reaction using HeLa genomic DNA and different fluorophore-labeled primers that target different guanosine sites in telomeric DNA. (**b**) The quantification result for the measurement of the amount of OG at telomeric DNA. Error bars are standard deviations from three independent experiments. The unpaired *t*-test was performed to evaluate the differences of extension rate. All *p* values were two-sided. (Reproduced from ref. [6] with permission from The Royal Society of Chemistry)

6. PCR products were purified by agarose gel and then subjected to Sanger sequencing (Fig. 4).

3.2.2 Sanger Sequencing Analysis of OG in Genomic DNA

1. 500 ng of HeLa genomic DNA was denatured and annealed with the corresponding extension primers (*VEGFA*-L, *TP53*-L, and *KRAS*-L, Table 3). The following procedures are the same as those for the analysis of synthesized DNA.
2. H_2O_2-treated HeLa genomic DNA was also used as the template for the sequencing analysis. The H_2O_2 treatment was performed by following a previously described method [24]. HeLa genomic DNA was incubated with H_2O_2 (400 μM), $CuCl_2$ (50 μM), and ascorbate (4.0 mM) in a PBS buffer (25 mM NaCl and 50 mM phosphate, pH 7.0) for 60 min at 25 °C (*see* **Note 5**).

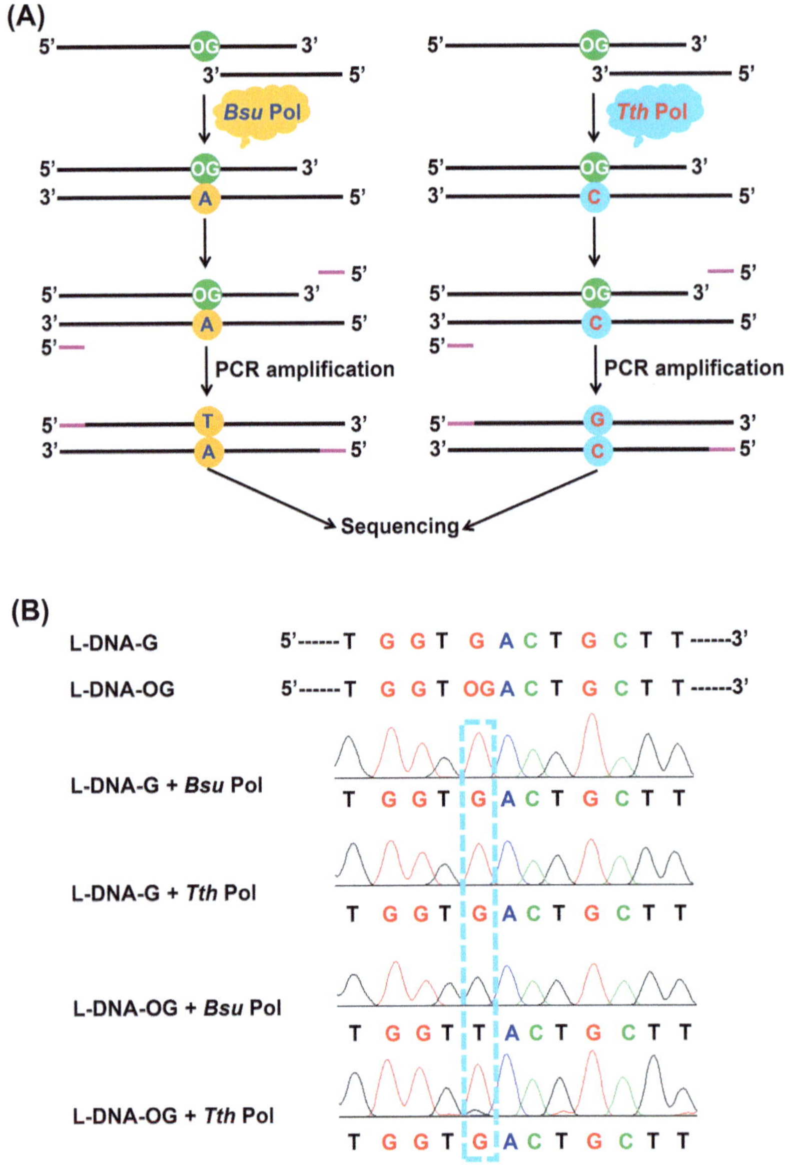

Fig. 4 Sequencing analysis of OG in DNA. (**a**) Schematic illustration of the strategy for the sequencing analysis of OG in DNA. (**b**) The results of sequencing analysis for the primer extension products by *Bsu* Pol or *Tth* Pol. (Reproduced from ref. [6] with permission from The Royal Society of Chemistry)

Table 3
PCR primers used for the sequencing analysis of OG in HeLa genomic DNA. (Reproduced from ref. [6] with permission from The Royal Society of Chemistry)

Gene/amplified region	Sequence (from 5′ to 3′)
VEGFA/chr6: 43772151–43772344	*VEGFA*-L (Primer-L): TGAGCAGATGTGTGACCACCGCCAC GACTTCTGACAGTGA
	VEGFA-F: TCTCCAGACCCTACCTCTGC
	VEGFA-R: TGAGCAGATGTGTGACCACC
TP53/chr17: 7686438–7686641	*TP53*-L (Primer-L): GAGCAGATGTGTGACGGCTAGCAGCTCAC TATTCACCCGA
	TP53-F: TGGGGCACACCATTCAAAGA
	TP53-R: GAGCAGATGTGTGACGGCTA
KRAS/chr12: 25248470–25248719	*KRAS*-L (Primer-L): GGTTAGAGCAGATGTGACGCCCTCC CAGCCCATGATCTTC
	KRAS-F: GAGGGGTCGTTAAGGCCAAA
	KRAS-R: GGTTAGAGCAGATGTGACGC

3. PCR products were purified by agarose gel and then subjected to Sanger sequencing (Fig. 5).

4 Notes

1. dNTP should be aliquoted to small size and stored at −20 °C no more than 3 months.
2. The extension time and dATP concentration should be carefully optimized due to the none-specific extension.
3. Deferoxamine mesylate was used as the oxidation inhibitor for protecting DNA from excessive oxidation during the whole DNA extraction process.
4. The excessive L-primer should be eliminated to avoid the primary template amplification.
5. All chemicals should be freshly dissolved in Milli-Q water in DNA oxidative assay.

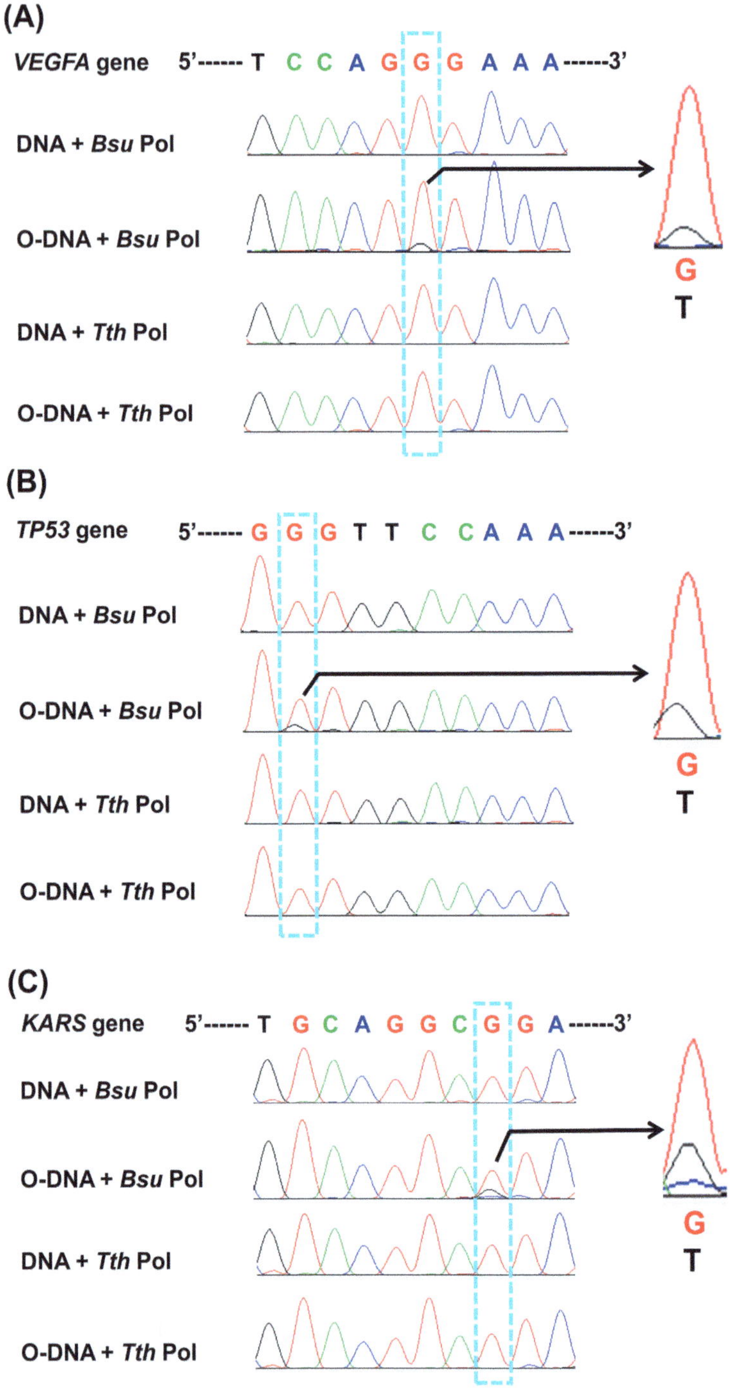

Fig. 5 Sanger sequencing analysis of OG in genomic DNA of HeLa cells. Three genes (*VEGFA* (**a**), *TP53* (**b**), and *KRAS* (**c**)) were amplified by using HeLa genomic DNA as the template. HeLa genomic DNA with or without H_2O_2 treatment was annealed with the corresponding extension primers and then extended by *Bsu* or *Tth* Pol. The PCR products were subjected to Sanger sequencing. O-DNA represents H_2O_2-treated HeLa genomic DNA. Shown in right are the corresponding peaks with G-to-T conversion. (Reproduced from ref. [6] with permission from The Royal Society of Chemistry)

Acknowledgments

The work is supported by the National Natural Science Foundation of China (21672166, 22074110, 21721005).

References

1. Forrester SJ, Kikuchi DS, Hernandes MS, Xu Q, Griendling KK (2018) Reactive oxygen species in metabolic and inflammatory signaling. Circ Res 122:877–902
2. Schumacker PT (2015) Reactive oxygen species in cancer: a dance with the devil. Cancer Cell 27:156–157
3. Genereux JC, Barton JK (2010) Mechanisms for DNA charge transport. Chem Rev 110:1642–1662
4. Alenko A, Fleming AM, Burrows CJ (2017) Reverse transcription past products of guanine oxidation in RNA leads to insertion of A and C opposite 8-Oxo-7,8-dihydroguanine and A and G opposite 5-guanidinohydantoin and spiroiminodihydantoin diastereomers. Biochemistry 56:5053–5064
5. Yuan BF, Jiang Y, Wang Y (2010) Efficient formation of the tandem thymine glycol/8-oxo-7,8-dihydroguanine lesion in isolated DNA and the mutagenic and cytotoxic properties of the tandem lesions in Escherichia coli cells. Chem Res Toxicol 23:11–19
6. Tang F, Liu S, Li QY, Yuan J, Li L, Wang Y, Yuan BF, Feng YQ (2019) Location analysis of 8-oxo-7,8-dihydroguanine in DNA by polymerase-mediated differential coding. Chem Sci 10:4272–4281
7. Liu T, Ma CJ, Yuan BF, Feng YQ (2018) Modificaomics: deciphering the functions of biomolecule modifications. Sci China Chem 61:381–392
8. Feng Y, Xie NB, Tao WB, Ding JH, You XJ, Ma CJ, Zhang X, Yi C, Zhou X, Yuan BF, Feng YQ (2020) Transformation of 5-carboxylcytosine to cytosine through C–C bond cleavage in human cells constitutes a novel pathway for DNA demethylation. CCS Chem 2:994–1008
9. Cheng QY, Xiong J, Ma CJ, Dai Y, Ding JH, Liu FL, Yuan BF, Feng YQ (2020) Chemical tagging for sensitive determination of uridine modifications in RNA. Chem Sci 11:1878–1891
10. Xiong J, Yuan BF, Feng YQ (2019) Mass spectrometry for investigating the effects of toxic metals on nucleic acid modifications. Chem Res Toxicol 32:808–819
11. Xiong J, Ye TT, Ma CJ, Cheng QY, Yuan BF, Feng YQ (2019) N6-Hydroxymethyladenine: a hydroxylation derivative of N6-methyladenine in genomic DNA of mammals. Nucleic Acids Res 47:1268–1277
12. Qi CB, Ding JH, Yuan BF, Feng YQ (2019) Analytical methods for locating modifications in nucleic acids. Chin Chem Lett 30:1618–1626
13. Lan MD, Yuan BF, Feng YQ (2019) Deciphering nucleic acid modifications by chemical derivatization-mass spectrometry analysis. Chin Chem Lett 30:1–6
14. Chen B, Yuan BF, Feng YQ (2019) Analytical methods for deciphering RNA modifications. Anal Chem 91:743–756
15. Chen B, Xiong J, Ding JH, Yuan BF, Feng YQ (2019) Analysis of the effects of Cr (VI) exposure on mRNA modifications. Chem Res Toxicol 32:2078–2085
16. Tubbs A, Nussenzweig A (2017) Endogenous DNA damage as a source of genomic instability in cancer. Cell 168(4):644–656
17. Fleming AM, Burrows CJ (2017) 8-Oxo-7,8-dihydroguanine, friend and foe: epigenetic-like regulator versus initiator of mutagenesis. DNA Repair (Amst) 56:75–83
18. Pan L, Zhu B, Hao W, Zeng X, Vlahopoulos SA, Hazra TK, Hegde ML, Radak Z, Bacsi A, Brasier AR, Ba X, Boldogh I (2016) Oxidized guanine base lesions function in 8-oxoguanine DNA glycosylase-1-mediated epigenetic regulation of nuclear factor kappaB-driven gene expression. J Biol Chem 291:25553–25566
19. Zarakowska E, Gackowski D, Foksinski M, Olinski R (2014) Are 8-oxoguanine (8-oxoGua) and 5-hydroxymethyluracil (5-hmUra) oxidatively damaged DNA bases or transcription (epigenetic) marks? Mutat Res Genet Toxicol Environ Mutagen 764-765:58–63
20. Fleming AM, Ding Y, Burrows CJ (2017) Oxidative DNA damage is epigenetic by regulating gene transcription via base excision repair. Proc Natl Acad Sci U S A 114:2604–2609
21. Yuan BF (2020) Assessment of DNA epigenetic modifications. Chem Res Toxicol 33:695–708

22. Wu WJ, Yang W, Tsai MD (2017) How DNA polymerases catalyse replication and repair with contrasting fidelity. Nat Rev Chem 1:1–16
23. Horhota A, Zou K, Ichida JK, Yu B, McLaughlin LW, Szostak JW, Chaput JC (2005) Kinetic analysis of an efficient DNA-dependent TNA polymerase. J Am Chem Soc 127:7427–7434
24. Gu C, Wang Y (2007) In vitro replication and thermodynamic studies of methylation and oxidation modifications of 6-thioguanine. Nucleic Acids Res 35:3693–3704

MIX
Papier aus verantwortungsvollen Quellen
Paper from responsible sources
FSC® C105338

If you have any concerns about our products,
you can contact us on
ProductSafety@springernature.com

In case Publisher is established outside the EU,
the EU authorized representative is:
Springer Nature Customer Service Center GmbH
Europaplatz 3, 69115 Heidelberg, Germany

Printed by Libri Plureos GmbH
in Hamburg, Germany